**IEE TELECOMMUNICATIONS SERIES 46**

Series Editors: Professor C. J. Hughes
J. J. O'Reilly
Professor G. White

# Intelligent Networks

## Principles and Applications

**Other volumes in this series:**

# Intelligent Networks

## Principles and Applications

John Anderson

The Institution of Electrical Engineers

Published by: The Institution of Electrical Engineers, London,
United Kingdom

© 2002: The Institution of Electrical Engineers

The Institution of Electrical Engineers,
Michael Faraday House,
Six Hills Way, Stevenage,
Herts. SG1 2AY, United Kingdom
www.iee.org

**British Library Cataloguing in Publication Data**

Anderson, J. R.
Intelligent networks: principles and applications.
(IEE telecommunications series; no.46)
1. Artifical intelligence  2. Computer networks
I. Title  II. Institution of Electrical Engineers
004.6

**ISBN 0 85296 977 5**


Typeset in the UK by Tradespools Ltd, Frome, Somerset
Printed in the UK by MPG Books, Bodmin, Cornwall

# Contents

# Preface

The purpose of this book is to give some practical insight into how public telecommunications networks have advanced from the old days of offering just simple telephony services to becoming today's 'intelligent' networks. Telephone companies now operate in an intensely competitive environment, and so they have to constantly struggle to find more ways to encourage customers to make ever more use of the installed network equipment. Network operators are therefore constantly looking for new services that will make the telephone system more attractive, easier to use and perhaps even more indispensable for its users. The overall goal is of course to increase the revenue-producing ability of the existing phone systems. The aim of the 'intelligent network' (IN) has always been to help the telephone companies achieve just this, but in a rigorously standard manner, so that network equipment from different suppliers can work together. IN seeks to provide the internal network rearrangements that give the agility and flexibility needed to bring almost any new services into life very quickly.

The intended audience for this book is really anyone who has an interest in this subject without necessarily yet having any significant level of expertise in intelligent networking. The background to IN is formal and complex, because its fundamental driver is a set of agreed interfaces between different functions provided in different vendors' computing and telecommunications equipment. This book therefore takes pains to give what is hoped to be an easily understandable account of IN for readers who are unfamiliar with the background detail.

A typical target reader might therefore be a student or engineer with background knowledge in computer hardware and software, but with little specialist experience in IN. The book will have succeeded in its aim if it enables such a reader to assimilate the essentials of the subject of IN with reasonable speed. Because of the potential complexity, we have aimed to keep the text reasonably economical, and to focus primarily on explanation and illustration of the principles rather than the presentation of comprehensive coverage. For further information we have in general referenced out to the plethora of background and specialist supporting documents that are available from the international and regional standards organisations and elsewhere.

Over the years the term 'intelligent network' has been used in association with a variety of meanings and nuances. This has inevitably brought into existence several different notions of what an 'intelligent' network actually is, so the first task for a book with IN in the title is to establish exactly what we are actually talking about. What actually do we mean by an intelligent network?

We live in a world of fast-advancing technology, and we are adapting to the various 'convenience' lifestyle aids that modern telecommunications equipment gives us. We are quickly becoming used to the availability of home-shopping through the Internet, automated voice-response bank clerks, the ability to be constantly in contact wherever we are on the planet, and future prospects of virtual reality holidays. It may actually be more pertinent to question what a 'non-intelligent' network might be! A casual Internet search on the phrase 'intelligent networking' yields information on an alarmingly wide array of widely disparate topics. These include, for example, 'intelligent' house wiring and burglar systems, greenhouse-sprinkling systems, neural road maps for chimpanzee brains and a host of interesting academic papers on network design theory. We therefore need to tie down our particular use of the phrase 'intelligent network' early in the book, and so we will build further on the outline definition of the aims of IN that we gave in the opening paragraph of this preface.

Many papers and standard recommendations have been written about the progression of the IN vision. However most major network operators have completed their IN implementations based on the standards laid down in the early 1990s. Even though the international standards have come a long way since the early days network operators with IN systems have generally not upgraded in line with the more advanced IN recommendations. New services are still often being introduced using the early well-tried IN standards. This book therefore concentrates more on the practical exploitation of the existing IN systems than on speculation of how newer IN standards might be used. We focus on the use of the existing IN systems for advanced services, illustrating the principles with practical examples.

Earlier chapters in this book concentrate on the internationally established standards for intelligent networking in fixed networks. The ITU-T IN 'capability sets'[1] establish the progression of development of the international standard for IN. Within this contextual scope, regional standards bodies, such as ETSI in Europe and ANSI in the USA, have been active in establishing the details of workable frameworks for the design of IN implementations.

Later parts of the book are about newer standards and developments because, whilst the introduction of new IN features using the initial systems in fixed networks has been reasonably steady, the use of IN in other areas, particularly mobile networks, is demonstrating fast take-up and rapid growth. Also, the IN designed 10 years ago was optimised for voice telephony, and big questions now

---

[1] The ITU-T (International Telecommunications Union — Telecommunications division), which was called CCITT (International Consultative Committee on Telephony and Telegraphy) until 1994, is the over-riding international standards body for telecommunications. The ITU-T has defined a series of IN functionality releases known as 'capability sets'.

are to do with how these principles carry forward into the new generation of multimedia and 'Internet' services.

One of the earlier pioneering texts on IN [1] stated that by the year 2000 the term 'intelligent network' would be history. Whilst the major standardisation work on IN was indeed completed in the last century, we have to remember that the standards formulation is just the beginning for a successful product! IN now survives because it has entered the fundamental realms of 'business-as-usual', and we are pleased to be able to report that IN is still very much alive. However, IN features do now tend to be taken for granted as part of the standard set of today's network switching functions, and so the use of the term IN as a visionary concept is probably now historical.

# Acknowledgments

This book has drawn heavily on practical experiences gained whilst launching IN services into the British Telecom UK domestic network. This was a large-scale team effort and much of the material in the book relates to work done in conjunction with many colleagues. In assembling this work I am particularly indebted to those co-travellers who provided generous assistance. The following colleagues kindly provided me with extra source material: David Gardner (BT), Richard Hennessy (Conformance Standards Ltd), John Shepherd (BT) and Richard Swale (BT). I am also indebted to Nigel Dadge (BT), Andy Fisk (BT), Bob Hinchliffe (BT), David Smith (mmO2), Graham Fitt (Conformance Standards Ltd) and Dick Cullingford (Conformance Standards Ltd) for their invaluable assistance at the technical review stages.

I would like to thank Charles Hughes, Gerard White and Roland Harwood at the IEE for driving the project through from beginning to end with patience and continuous encouragement, and Richard Swale again for organising the final editorial process that brought the book into existence. However, top marks for unlimited support and long-suffering patience throughout the project go to my family, particularly to my wife Elizabeth – who gave up many hours to help me with the proof-reading and organisation of the drafts.

# Chapter 1
# Introduction to intelligent networks

*The past is a foreign country: they do things differently there.*
                    Hartley, L.P.: 'The Go-Between' (Penguin Classics, 1953)

## 1.1 The basics of intelligent networks

The world of telecommunications has changed dramatically since the 'intelligent network' was defined, around 20 years ago. However, the early principles, which were established at that time, have now become embodied in most public telephone networks and are the driving force behind many advanced telecommunications services.

### 1.1.1 Origins

Like many things, intelligence in telephone networks is not new. In many ways, public networks were at their most intelligent perhaps 70 years ago, when every subscriber had a local telephone operator, perhaps at the local shop, who knew all her customers by name and probably in person as well. 'Calling by name' was a natural service – you would ask to be connected to a particular person and the operator would endeavour to track that person down and connect you to them. If you asked her for the florist shop or the taxi service, she probably wouldn't have had to trouble you for the number. A service such as 'calling name delivery' was naturally included for local calls – the operator would announce who the caller was when she put the call through to you.

Call diversion (e.g. to the doctor currently on duty) was easy – the operator would have the doctor's current whereabouts on a piece of paper in front of her, if she didn't know it already, and she would put you through to wherever he was currently located. Billing was done by writing out a ticket for the call, so 'alternate billing' or 'reverse-charge calls' were very straightforward services – she would just write down on the ticket the number, or the name of the person to be charged. A 'conference call' service was possible. Early switchboard equipment could cross-connect several simultaneous calls, and the operator could offer her services in supervising the conference. Most of the services we now call 'advanced' were feasible from the manual boards of 70 years ago. Tele-shopping was fairly common, because the telephone operator was usually also the local post-mistress and grocer. I remember our own post-mistress in Oxfordshire still provided a tele-shopping service in the 1950s – she used to deliver groceries in her 1932 Austin 7 Ruby a few hours after receiving the telephoned order.

However, the days of ubiquitous operator control were soon to be numbered when some Kansas City undertakers tried to cheat a rival undertaker by installing an efficient incoming call screening service on his exchange lines. They bribed the local switchboard operator to divert his incoming business calls to their own establishments rather than his. However, they hadn't reckoned on the inventiveness of their victim, whose name was Almon B. Strowger. When he discovered what was going on, Strowger set out to redress the balance. He invented an automatic call-switching device, which was destined to revolutionise the telecommunications industry. This device, the construction of which was based on his top hat, was called a 2-motion selector. It enabled one electrical inlet to be connected to any one of 100 outlets on receipt of a pair of pulsed digits. When two sets of digit pulses from a mechanical dial arrived at his new switch, the first train of pulses would step the vertical motor of his selector up to the required level and the second train of pulses would step the rotary selector around to the required horizontal outlet. The operator's manual switching function could thereby be dispensed with and calls routed automatically to their proper destination without unlawful diversion along the way.

The invention of this ingenious device in 1888 heralded the advent of automatic telephony, and the first public automatic exchange was opened in 1892 at La Porte, Indiana, USA [2]. Large armies of switchboard operators were no longer needed and network traffic growth rocketed. However, whilst automation increased, 'intelligence' in the network, in terms of the variety and range of personal services available to customers, gradually diminished (Figure 1.1) with the demise of the hitherto ubiquitous switchboard operator.

The 1920s, 1930s and 1940s were periods of intense inventive activity, brought on partly by military necessity. Many ingenious devices were invented to enhance the automation of telephony – notably the electromechanical 'register translator' which could translate dialled digits into routing digits. This greatly enhanced the embedded 'intelligence' of the network, leading to the automation of trunk calls in the early 1950s.

After this followed the concept of centralised control of telephone exchanges, and 'stored program control' (where pre-programmed network routings were held) introduced more captive intelligence. These innovations have developed into today's digital exchanges, which use fast common channel inter-exchange signalling to provide complex network services.

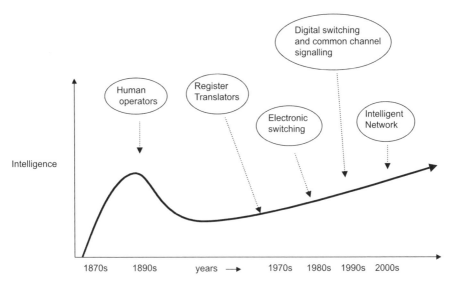

*Figure 1.1    The rise and fall of intelligence in networks*

For a long time, therefore, the central theme of 'intelligence' in networks has been to do with translating a dialled digit string into a different string of digits, forming an address routing. The obvious illustration is the Freephone service, where a caller dials an advertised number which is then translated by the network into a routing number, and the particular translated number can vary depending on factors such as time of day, location of origin of the call, etc. Similar principles apply to 'virtual private network' calls, where the translated digits are also dependent on the calling number. In countries such as the USA, where it is still commonplace for telephones to have letters as well as numbers on the dials and key-pads, there is far greater opportunity for advertisers to catch the public imagination with a string of letters instead of numbers. An address such as 1800 CARPET tends to be rather more memorable, and thereby revenue producing, than the equivalent digit string – 1800 116628. The potential high revenue generation of these sorts of calls is of course one of the main reasons for the high interest in intelligent network services.

The term 'intelligent networks' (IN) was introduced by Bellcore in the USA in the mid 1980s as a consequence of the upheaval following the regulator's break-up of AT&T's monopoly hold over the US telecommunications industry. Bellcore, which has since been renamed as 'Telcordia Technologies', was the United States telecommunications research body that was jointly funded by the seven regional Bell operating companies (RBOCs). The RBOCs at the time were also keen to provide the lucrative new additional services such as Freephone and Alternate Billing, which had proved so successful for AT&T. The RBOCs therefore commissioned Bellcore to develop the standard, which Bellcore called 'intelligent networks'. Consequently Bellcore led the way with IN standards for the latter part of the 1980s. We discuss this early development of the IN standards in the US at greater length later in the book.

Whilst many definitions now abound for the catch-phrase 'intelligent network', technically the original underlying principle was that of separation of the logic controlling services from the 'basic' call control function in existing telephone exchanges. Telephone exchanges have of course always been 'intelligent' to some degree, even if this just refers to an ability to translate a dialled digit string to an outgoing routing number, or port number associated with a customer's line. As well as such translation facilities the current generation of digital exchanges has standard facilities for services such as abbreviated dialling and call waiting, using 'embedded' service logic (Figure 1.2). However, the original essence of the 'intelligent network' standards was to separate the software for new services from the basic call control software and to use a standardised interface between the two.

These pre-existing embedded services allow operators to achieve some degree of competitive differentiation, but for new switch-based services there is a heavy dependency on the switch manufacturer's ability to provide these at a cost and time-scale which is suited to the operating company's commercial limitations and opportunities. An appreciation of this commercial dilemma gives an insight into the operating companies' major interest in intelligent networks. The attraction is the separation of the service logic into a separately controllable entity, where services can be provided, modified and controlled by the operator rather than the switch manufacturer (Figure 1.3).

### 1.1.2 Extracting the service logic

The switches in Figure 1.3, which have been upgraded to incorporate the IN 'trigger' functions, are renamed as intelligent network 'service switching points' (SSPs). An SSP 'triggers' when it detects a pre-set event during a call, such as a particular string of digits dialled, or a particular line going off-hook, and it sends a request message to a Service Control Point (SCP). This extra capability, which transforms a telephone exchange into an IN SSP, acts as a sort of 'go-between', i.e.

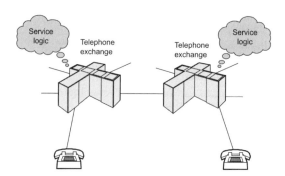

*Figure 1.2   'Embedded' service logic*

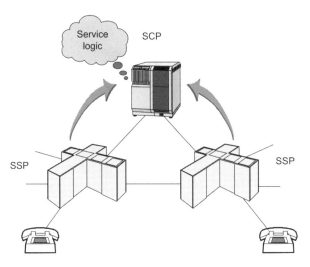

*Figure 1.3* *'Separated' service logic*

a junction between the world of the transport network's bearer circuit control and the service logic running in the SCP.

The SCP's function is to execute the service logic needed to implement the required service. The SCP is shown, for simplicity, as a single node. It almost certainly will have access to an internal or external database. In practice the SCP may be a composite node, made up of several constituents, and the data, or service logic, or both, can exist in a geographically distributed architecture.

### 1.1.3 Interactive voice functions

Many intelligent network services need the assistance of peripheral devices, such as automatic voice announcements tailored for specific service situations, under the control of the separated service logic (Figure 1.4). These devices are known as 'intelligent peripherals' (IPs).

An IN IP is a platform that supplements the transport layer's abilities to conduct dialogues with end-users. It acts as an interpreter between the SCP and the user, enabling the service logic to communicate with the service user in order to obtain more information than is passed across in the initial request (a set of dialled digits) from the caller. Procedures have been standardised which enable an SCP to instruct an IP to conduct a dialogue with a user and return the required information to the SCP. This dialogue uses automatic voice guidance, requesting a response from the caller in the form of a vocal reply, which can be recognised automatically, or in the form of DTMF (dual tone multi-frequency) digits, keyed on the key-pad of the telephone.

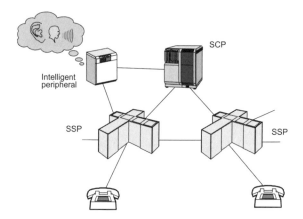

*Figure 1.4   Automatic voice announcements included*

### 1.1.4 The 'classical' IN architecture and its elements

Figure 1.5 shows a block diagram of the main network elements of the traditional intelligent network architecture. It is often called 'classical' because its overall structure has changed little over the past 15 to 20 years, and it is still relevant as a guiding reference to the framework of the IN vision, which has endured throughout.

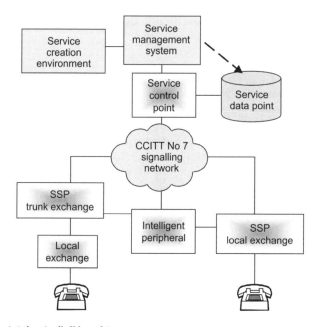

*Figure 1.5   A 'classical' IN architecture*

Central to the architecture is the SS7 common channel signalling system [3]. SS7, sometimes known by the former title 'C7' (CCITT No. 7) was designed for the transport of telephony call set-up messages, but it has been modified, by the addition of non-circuit-related signalling capabilities, to carry IN messages between the IN elements shown in Figure 1.5. With these modifications the SS7 signalling system effectively provides the 'glue' that holds the IN elements together, and in many respects is the most important intelligence component of all. The enhancements needed to the SS7 specification to achieve this are the addition of the signalling connection control part (SCCP) and transaction capabilities application part (TCAP). These functions sit alongside the telephony user parts, using the same underlying signalling data link capability. TCAP running over SCCP supports a computer–computer communication framework and an IN messaging protocol, known as intelligent network application protocol (INAP).

The SS7 network contains much distributed 'intelligence', both in terms of information about the availability of the rest of the nodes in the network and in terms of flexible 'global title' routing tables positioned at intermediate locations in the network. The SS7 network enables the SSPs and IPs to communicate with the SCPs reliably and in real-time. These aspects of signalling intelligence are discussed further in Chapter 3.

Figure 1.5 also shows the IN service management system (SMS) and service creation environment (SCE). A new service is developed, assembled and tested using an IN SCE before being down-loaded, via the SMS to the various SCPs. Typically SCEs use software components known as service independent building blocks (SIBs), with graphical tools for linking and activating them. Service and user data are stored in the service data point (SDP), and there are standards that allow this network element to be remote from the SCP, although for convenience the data are often held in the SCP itself. However, a separate SDP allows data to be centralised and accessed by several SCPs. It also allows the SCP to access data in a different network, and this could be in a service provider's, or customer's, system.

As well as organising the distribution and download of the new services, the SMS also controls the network data population issues for new and existing services. Service data is needed in the SSPs as well as the SCPs, although the former is likely to use existing OSS (operational support systems) mechanisms for managing switch data. Figure 1.5 only really shows the new elements for an IN-enabled network. Most of the essential day-to-day background support functions, such as customer provisioning, billing, performance monitoring and network management, all of which need to be upgraded for new IN features, are not shown in Figure 1.5.

### 1.1.5 Locating the 'intelligence'

A recurring point of interest with the choice of architectural arrangements is where to put the intelligence. Before IN, network 'intelligence' was dispersed around the network, and the call control was distributed along the speech path, and did not divert to seek out specialised points of intelligence. For instance, number translation facilities were usually provided at appropriate points along the 'natural' call chain, and the format of the network numbering scheme used to dictate the physical path taken by the call.

However, with greater centralisation, the 'intelligence' tended to become more concentrated and sophisticated. There was a consequential drive to provide the intelligence in central locations, making for easier management than dispersed intelligence schemes.

With centralised intelligence the service logic and data are usually physically separate from the preferred speech path of the calls and we need to bring the two into communication. As we have described, the standard IN method is to use remote signalling to communicate between the switch and the central intelligence. However, there are some circumstances where it is more practical to actually route the call itself to a point of intelligence and then onward route to the required destination. This is known as the 'service node' (SN) concept, where the transport switching control and the intelligence centre are combined in a single specialist node. SNs are sometimes referred to as 'point' solutions, or even 'IN-in-a-box' solutions, and are sometimes used to enable IN services to be provided quickly and cheaply. The disadvantages are firstly that such solutions are not good for scalability and secondly that they are provided on the call speech path. This means that the speech circuits will almost certainly have to traverse greater distances, as the calls now have to 'trombone' into and out of the SN before reaching the ultimate call destination. With a long distance call this can lead to the need for complicated network designs involving echo-cancellation devices as the transmission propagation delay becomes noticeable. The SN concept becomes attractive in situations where IN triggering functions do not yet exist. In this case an option for fast introduction of a Freephone service (for instance) is to introduce a single specialist node, or sub-network of nodes, which either include service control functions, or are capable of triggering to an external SCP. As we have said, this leads to speech-path inefficiencies, but this solution may be feasible for an initial low level of demand for a start-up service. This is sometimes known as 'overlaying' IN.

As suggested earlier, another commonly deployed 'interim' solution is the use of a network node known as a 'service switching control point' (SSCP). In this node an upgraded digital exchange is provided with an ability to break out from normal call processing to an internal service logic module which behaves in a similar way as an external SCP. There is an obvious performance advantage, because fast inter-process messaging methods can be used instead of an external messaging system such as SS7. A disadvantage is the increased data distribution task – data changed centrally need to be relayed to the internal database on the SSCP. If the required service logic is not locally resident then the node contains normal SSP functions and so can trigger out to an external SCP for assistance in special cases.

The SSCP approach is understandably favoured by established equipment vendors, because SSCPs offer the benefits of IN solutions to operators whilst still broadening the market product range for the vendor. SSCPs in fact represent one of the earliest ways of implementing IN. They were actually used on occasion as dedicated service nodes in central points in smaller networks.

### 1.1.6 What is an 'intelligent' network?

This chapter reviews some of the historical origins and concepts behind the Intelligent Network 'vision' whilst continuing to introduce some of the network elements such as service control points and service switching points, which are needed in the embodiment of this vision. We also discuss some aspects of introducing IN principles to existing digital switching networks, but first we must define what we really mean by IN.

In the quest for the basic principles of IN we constantly return to this question of what IN actually is. Whilst the principles have held, the intelligent network concept has appeared in a variety of guises over the past 15 to 20 years. Also, because the term was held in high regard in business cases, IN has sometimes been over-used as an attention seeker. In fact the 'de facto' definition of the vision has broadened so that the range of meanings now attributed to the term is so wide that the definition of IN is a somewhat daunting task! However, in the main we restrict ourselves to the original meaning of IN, which is the separation of service logic from bearer control in telephone exchanges.

From a service viewpoint, the aim should be to organise services around the users, rather than the other way round. A perceptive early articulation of the core concept of the IN was 'the creation of highly functional and highly customer-specific services through software manipulation in a multi-vendor, multi-network, environment [4].

With the recent meteoric rise in popularity of mobile telephony, another viewpoint insists that, with a modern popular desire for people to be always in contact, regardless of where they might be physically located, the most notable capability of IN is that of universal mobility. The notion of IN has naturally expanded to embrace the idea of intelligence 'everywhere', which is a theme that we return to in later chapters.

### 1.1.7 The role of international standards

It is sometimes said that the advantage of standards is that there are so many of them to choose from! Obviously this alludes to the problem of competing standards and the dilemma that equipment manufacturers find themselves in when trying to design their product range in advance of their competitors. How does one know to which emerging standard to attach one's allegiance? This is obviously a big problem, and one that can have major financial consequences either way.

The international standards organisation surrounding the telecommunications industry is vast – there is a large hierarchy of study groups, working parties and committees. The input and output reports of these groups are formidable in their volume, depth and detail. Decisions made at standards meetings can substantially affect the major investment programs of the industry's equipment manufacturers, network suppliers and the operating companies, all of whom therefore take part in the international standards debates with the intention of influencing the outcomes towards their preferred directions. Consequently, much of the expertise in the fine details of a complex topic such as intelligent networks resides within these international standards working groups and sub-committees. The big challenge for the industries supporting the standard-making process is to balance their investment in this natural drive towards greater standards complexity against the practical usefulness of the outcome of the process in their own organisations. A common practical issue is the fact that the people who attend the standards working groups are not always the same people who are tasked with solution implementation. Both functions are usually full-time occupations, and the two disciplines need to find time to communicate, which is not always as easy as it might seem.

We discuss the various standards bodies in Chapter 4, but in general, with our detailed examples of working IN solutions, we adhere to the ETSI Core INAP variant of the ITU-T Capability Set 1. This is because of the popularity of this variant, which is now used widely in Europe and elsewhere. It is also the case that, because of its relationship with the wider ITU-T standard [6], it shares common roots with the US AIN standard and Asian standards. However, the aim of this book is really to introduce the concept of IN, illustrating basic principles without extensive comprehensive coverage of the detail. We therefore explain the terminology we need to achieve this, but will not attempt to explain the larger part of the detail which is to be found in the standard documents themselves.

The reason for the heavy emphasis on international standards for IN is that digital switch technology has been developed over the past 30 years largely on a proprietary and competitive basis. Different vendors and operators have worked to provide the features they individually regard as important in their digital switches and their controlling software suites. Therefore there has been little commonality of internal structure, messaging, control methods, storage and retrieval systems and suchlike, so there has been little scope for standardisation on the internal architectural arrangements inside the switches.

Digital telephone exchange systems are still amongst the largest and most complex real-time software systems in the world. To compound the complexity, these designs have been (and still are) subject to regular modifications and upgrades in response to intense commercial and regulatory pressures to provide new service features. Consequently there are in existence many huge and monolithic telephony switching systems, perhaps in the later years of their life cycles, with programs which have been implemented in out-of-date programming languages and which have been substantially modified, or patched, along the way. With many modifications being made to these large systems, which were not originally designed for the scale of upgrade to which they have been subjected, great care is needed to ensure that maintainability is not compromised [5].

For this reason a large-scale overhaul and upheaval of the fundamental call processing software in order to introduce the complexity needed for the IN breakout points was necessary. The network operators viewed this with trepidation. In practice, the upgrades have generally been introduced gradually for this reason. This goes some of the way to explain the slowness that has characterised the introduction of IN features in the networks over the past 10 years. New builds on switch software might be loaded every 6 or 12 months, and a limited amount of IN triggering might be introduced each time. The amount has often been limited by financial factors – usually business case justification is needed, in terms of new revenue-producing service predictions, for each step along the way. The amount is also limited by practicalities on how much change a new software build can contain. IN features will be included alongside many other (non-IN) changes, bug fixes, etc., and there has to be a limit on how much change an operational switch can absorb without putting normal operation at risk. The period between builds is usually unhurried for operational reasons – often a new build needs several months to 'settle down'. Inevitably there will be a certain amount of unexpected behaviour in a new build, and time is needed for maintenance routines to adjust to this.

## 1.2 Why did we need IN?

*We don't like their sound, and guitar music is on the way out.*
Decca Recording Co. rejecting the Beatles, 1962

### 1.2.1 The commercial drivers

From the network operator's viewpoint, the sort of advantages promised by IN which caught the imagination were:
- vendor independence
- fast service deployment
- customised services
- customer control of service data.

### 1.2.1.1 Vendor independence
After the major network operators had invested heavily in different varieties of new digital networks of SPC exchanges in the early 1980s they began to discover that introducing new service features was an expensive business. The industry entered a phase where network upgrades were slow and expensive, and the operators had to gear up for tortuous contract negotiations with their vendors, requiring the production of water-tight specifications and rigorous test schedules before a new service could be introduced for public service. Equipment vendors were generally in a dominant position because the operator had no choice but to obtain expensive proprietary network solutions from the suppliers of their existing network equipment. Also, large amounts of patience were often needed. It sometimes took several years to negotiate and specify a new service, capture the detailed

requirements, generate the design details and complete the testing. Then the new service had to compete with other upgrades and features to be included in the planned build release cycle.

Against this background, the prospect for the network operators of a new way of introducing new services independently of the established process was very attractive.

### 1.2.1.2 Fast and flexible service deployment

IN held the promise of a network operator inventing its own services, developing testing and integrating them, and quickly rolling them out into the network, perhaps just a few months after their conception. Without IN systems this could take several years, by which time the original driver for the service might have disappeared, resulting in the loss of a marketing opportunity.

Hence a generic network enhancement, such as IN, which might help end the regime of service-specific network enhancements, was viewed with great interest from operators' business and marketing viewpoints. The procurement of facilities for 'in-house' development of services using 'service creation' functions (introduced earlier in this chapter in discussion of the SCE) made good business sense.

Similarly, the opportunity for faster and cheaper methods of service introduction meant that there was more chance of catching fleeting 'windows of opportunity' for specialist services. For instance a voting service could be developed and installed ready to run alongside a popular television programme.

As the process of service introduction became cheaper using IN features, it became possible to be more experimental and inventive with new service ideas. If a particular service IN does not generate sufficient revenue it can be disabled and removed without heavy financial losses. In any case, services can be launched as trials using IN, so that their potential demand can be assessed, before full-scale introduction.

A basic characteristic of IN is the re-use of generic software functions for different service scenarios, yielding obvious economies of scale. For example, a 'number translation' generic function will be re-used in different guises again and again in a multi-service IN.

### 1.2.1.3 Customised services

The notion of a service creation environment (SCE) under the control of the network or service provider, rather than the network supplier, became a fundamental raft of the IN concept. A versatile SCE would enable the service administrator to tailor services for particular customers. For instance, a user might like a particular variant of a VPN (virtual private network) service whereby his home phone is part of a VPN group during daytime business hours, but then would be reverted to a normal private line in the evenings and at weekends. In principle this sort of customisation is straightforward when the IN holds a service profile database for the user. The SCP database would be checked for every out-going call anyway, so the extra feature of a 'time-of-day' routing plan is a relatively small incremental feature.

Ultimately one can envisage that a customer's set of services could be provided by logic and data that are tailored to the customer's own set of requirements. This implies a movement towards a 'customer-centric' rather than 'service-centric' approach where the network operator, or the service provider, sells sets of service features tailored to particular users. In other words, service logic and data provision revolve around the customer's own service profile rather than a set of stand-alone service logic programs (SLPs).

### 1.2.1.4 Customer control of service data

A natural extension of customised services is to allow customers to have direct access to a sub-set of their service data. This is commonly provided by allowing users to have Internet access to a 'diary' profile in order to update their current locations. Alternatively, the data could reside on the customer's, or a third party's, computer and access provided from the network operator's SCP.

### 1.2.2 IN as a universal integrator

A recent perspective is that IN might be the panacea that integrates all networks, and incorporates all surrounding and 'off-network' intelligence. This is noticeable where broadband architectures are concerned, where network intelligence has a role to play with the background management tasks, such as linking session set-ups with available bandwidth.

It is important to understand that IN features are added as enhancements to networks that are not entirely stupid in the first place. Traditional telecommunications networks already have a certain amount of automatic 'intelligence' by virtue of their signalling systems, by virtue of their switch-based services provided by the switching equipment supplier and by virtue of ISDN features. It is useful therefore to assess the position before the IN enhancements, to understand how existing 'intelligent' features work before adding in the IN features. This is particularly important because there is a set of potentially complex problems, known as 'feature interactions', which are to do with IN and non-IN features working together and maybe against each other. To analyse and deal with these interactions it is necessary to carefully study the origins and mechanics of both. For this reason we spend some time (in Chapter 3) looking at the 'intelligence' in telecommunications that is not actually provided by IN. In fact IN upgrades could not operate without this 'prior' intelligence in public switched telephone networks (PSTNs).

### 1.2.2.1 Some other technical characteristics

We previously listed and described some of the commercial drivers for the IN vision; IN has been driven as much by pragmatic network possibilities as by commercial requirements. We have described the essential features of the IN architecture; we now summarise some of the other technical features that have characterised IN development:

- The construction of IN services has been based on the concept of generic software modules, which we have already introduced as 'service independent building blocks' (SIBs). Whilst there is valuable guidance on the use of SIBs in the ITU-T standards, this is really an 'off-shoot' from the main purpose of the standards deliberations, which was the definition of a standard real-time interface between the service switching software in the exchanges and the separated call control functions. Hence, whilst there are many 'service creation' systems, there are no over-riding standard recommendations for the implementation details or the interface with other IN elements. SCEs in use today are therefore fundamentally proprietary to the different manufacturers.
- GSM (global system for mobile communications) radio networks use what is essentially an application-specific IN for mobility management and for customer-controlled call-management functions. This is discussed, as well as the specific IN enhancements for the CAMEL (customised application of mobile networks enhanced logic) system in later chapters.
- Service logic independence from underlying network technologies is provided by 'middleware' application programming interfaces. The IN aims to provide a service-oriented network view, with the intention of concealing bearer-level practicalities from the service level as far as possible.
- IN is a 'coming together' of the computing world and the traditional telecommunications worlds. It makes use of 'open' computing interfaces, using results from computer technology. An example of this is the 'telecommunications information network architecture' (TINA), which is an object-oriented architecture, based on DPE (distributed processing environment) technology. In fact IN can be viewed as presenting a generic API (application programming interface) for network-transparent service provision, turning the network into a 'programmable entity' which integrates different connection network technologies. This is discussed in Chapter 6.
- As discussed earlier, much of the 'intelligence' of an IN is in fact provided by the SS7 signalling system. For instance there are translation tables with the signalling system (global title translations) which direct messages to appropriate IN SCPs. Some Freephone implementations use the dialled number itself as a global title and the translation tables can be updated in real-time to ensure that IN queries always go to the appropriate SCP node. This is discussed further in Chapter 3.

*Chapter 2*
# The foundations of IN

*Drill for oil? You mean drill into the ground to try and find oil? You're crazy.*
Drillers who Edwin L. Drake tried to enlist to his
project to drill for oil in 1859.

## 2.1 The service switching function

As Chapter 1 intimated, the term 'triggering' in intelligent networks refers to the process of breaking out from normal call-control procedures in digital telephone exchanges to remote service logic. Triggering is fundamental because the original purpose of the IN was to enable call control software to recognise pre-assigned events and conditions and suspend call processing whilst it obtained further call processing instructions from an external service control point.

However, when IN was introduced, the software which handles this 'breaking out' process could not simply be added alongside existing call processing functions. It had to be built in as an integral part of the basic switch call control. Because different telephone exchange manufacturers used different design philosophies for their call control, achieving a standard for the 'break-out' procedures was difficult. In order to achieve progress in building the standard recommendations for enhancing telephone networks with IN features, it was necessary to introduce a hypothetical modelling device known as the 'service switching function' (SSF) to the telephone exchanges.

The illusion was created of an apparent 'separation' between the new IN access function (the SSF) and the existing call control function (CCF). In fact, because the introduction of IN triggers could only be achieved by fundamental modifications to existing call processing software, including the 'embedding' of the 'break-out' points in the CCF, there could not be a real functional split to correspond to this modelling construct. For this reason, the specification of a standard interface between the CCF and the SSF was not possible.

Although this issue caused some confusion early on (particularly when regulators made attempts to enforce an open CCF – SSF standard), it did enable the modelling process to progress to the completion of the development of a set of workable IN standards.

This theoretical split between CCF and SSF is illustrated in Figure 2.1, which shows a separation between the bearer control and the basic call process. The bearer control is to do with the handling of the circuit-switched transport connection, including the seizure and release of circuits. The basic call process is concerned with functions such as the handling of address digits, call routing, and any extra facilities provided, such as call forwarding or the presentation of CLI (calling line identification) information.

Traditionally a single circuit-related network signalling messaging protocol has provided both bearer control and call control signalling. This signalling protocol is commonly ISUP (ISDN user part of SS7) [7]. However, bearer independent call control (BICC) standards are becoming available that will enable the functions to be handled separately [8].

The 'customer data' function in Figure 2.1 refers to switch-based information held against particular lines. This could be data to do with the states of switch-based services (such as call barring and call waiting) but it can also contain information about addresses to be used during call set-ups. For instance, a 'presentation' number may be held against a line if a customer wants something other than the network number of the calling line to be displayed at the called end's terminal for the CLI service. For incoming calls, a line may be on diversion to another number, in which case the diverted number would be stored in the customer data. This is essentially 'non-IN' data; it is data used for services provided by the switches themselves without any IN assistance.

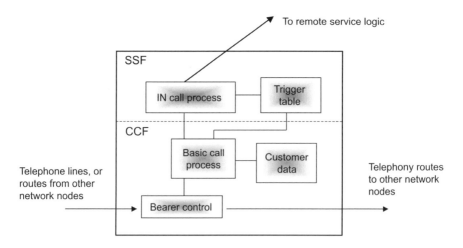

*Figure 2.1   The CCF/SSF 'split' in the SSP*

Therefore, services such as 'call forwarding' can either be 'embedded' in the switch call-control software or provided using IN features. However, the IN features have to be aware of any related switch-based services and take account of them. This is a facet of the 'feature interaction' issue, which was introduced briefly in Chapter 1 and is discussed further in this chapter as well as at later stages in the book.

Once triggering has occurred, the 'IN call process' function above the dotted line in Figure 2.1 supports any dialogues required with external service logic. The 'trigger table' is accessed from the CCF to determine the conditions in force for a particular trigger to occur, and it is also accessed from the IN call process to determine which switch-resident data elements should be included in messages sent to the remote service logic.

## 2.2  Triggering to remote service logic

The trigger (or 'break-out') points are implanted at strategic points in the basic call control process. This then provides the opportunity for dialogue to take place between the exchange and the control point that hosts the service logic.

### 2.2.1  Introduction to triggering

Once a trigger has occurred, there is a repertoire of 26 different INAP operations in ETSI Core INAP[2] that can be used for the communications between an SSF and an SCF. The term 'operation' is used to mean either a request made by one end for action to be taken at the other end, or to provide a response to a previous request. There are five INAP operations available for communications between an SRF and an SCF. The latter are used to obtain or provide further call-related information, usually from the caller. This can be achieved, for example, by playing automatic announcements and menu-driven invitations for the caller to respond verbally, or by pressing handset keys to generate DTMF signals. The INAP operations are listed later in Table 2.3.

Most CS-1 IN services are invoked by a 'trigger' event in the SSF, and an InitialDetectionPoint (I_DP) operation is used to convey information about the trigger event to the SCF. This information is carried in parameters, and the I_DP operation has parameters such as service key (a numerical reference to the set of trigger criteria that has been met by this particular trigger), address digits and a trigger detection point number. The ETSI Core INAP protocol specifies the service key as mandatory, and 16 other parameters as optional. The parameters are listed

---

[2] **Note**: The relationship between the various standards is explored in Chapter 4, but for practical purposes of explanation of IN principles we use the ETSI core INAP specification [10]. As we discussed in Chapter 1, ETSI core INAP is representative, closely related to ITU-T CS-1, and has proved to be a widely popular basis for IN implementations to date.

later in Table 2.2. Some of these parameters will nearly always be sent in any I_DP (such as the called number – and probably the CLI) but others (such as the ISDN bearer capability) will only be needed for more specialist services. It is the service key that indicates to the SCP which particular service logic program is required to run this service.

A simple example of an IN trigger is illustrated in Figure 2.2. This is a 'dialled digit' trigger (formally known in the CS-1 standard [9] as the 'analysedInforma-tion'[3] [DP 3] trigger detection point) and is in practice the most commonly used IN trigger. There are many variations possible. It could be set to be an identifiable dialling prefix, such as 080x (indicating a Freephone service in the UK) as shown here. It could equally have been set for 'digits '080815' or just '08'. The request message (the I_DP message) to the SCF could be sent immediately these digits have been received, or it could be delayed until the remaining digits have been received from the calling user.

It is often the case that the complete number length is unknown by the triggering SSP. This is because many networks do not have a standard number of digits in their customers' telephone addresses and it is often only the destination parent local exchange software that knows exactly how many digits are needed to uniquely identify the called user or, in this case, the called service.

In the example of Figure 2.2, a minimum number length, say eight digits, can be set in the data so that the I_DP could be sent to the SCF when the first eight digits ('08081570') have been received by the call processing software. In this case the SCP service logic may have information on the numbering scheme digit length, and, after analysing the first eight digits it may know that just two more digits are needed. It is able therefore to reply to the SSP with a CollectInformation message (described in Section 2.5.1) requesting the two remaining digits and the IN request can be processed when the SCP has the complete dialled address string.

However, if the digit length is unknown either in the SSP or SCP, then the network must normally wait for a period of time (known as the 'inter-digit time-out', which is typically 4 to 6 seconds) in case another digit should arrive. Another

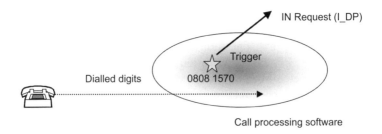

*Figure 2.2   A dialled-digit trigger*

---

[3] This is the ASN/1 (abstract syntax notation number 1) name for the analysed information trigger. More information on ASN/1 is given in Section 2.2.5.3.

way of avoiding this delay, which is used in some network implementations, is to allow the user to key a delimiter digit (e.g. a '#') to inform the network that the dialling phase is complete.

Having received all the digits, the SCF will typically return an instruction (the 'Connect' message, which is described in Section 2.5.1) to connect the call onwards to a translated number supplied by the SCF. This translated number, which is a network routing address, might for instance depend on the time of day, the caller's line identity (CLI) or a host of other factors. It could be that part of the service algorithm is to prompt the caller to enter an authorisation code using the DTMF key-pad on the phone. The routing destination might then depend on the code entered at this point.

Another common IN trigger is illustrated in Figure 2.3. This is the 'off-hook' trigger (formally known in the ITU-T CS-1 standard [9] as the 'origAttemptAuthorized' [DP 1] trigger detection point). This trigger causes the IN request (I_DP) message to be sent to the SCF when a user picks up the phone handset. This trigger could be set, for example, for a phone where the account has been discontinued because the previous tenant or owner has vacated the premises. The phone company might want to keep the line connected, hoping to attract the custom of the next occupant of the premises. An 'off-hook' trigger could therefore be set so that when the new incumbent picks up the phone handset they are connected to an announcement machine providing a welcome message.

In this case the SCP would reply to the I_DP with a message containing a 'Play Announcement' operation (Section 2.5.1), which instructs the SSP to connect the caller to an internal announcement machine, providing a reference number for the particular announcement to be played.

In a refinement of this service the SCF might use a 'Prompt and Collect User Information' operation (Section 2.5.1) to invite the caller to respond to an automated dialogue, perhaps by entering a digit on the key-pad, or by responding verbally. This mechanism is illustrated later in this chapter. The IN functional entity that is used to play the announcements and conduct user-dialogues is an IN

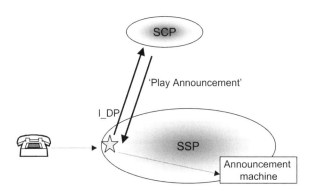

*Figure 2.3   An 'off-hook' trigger*

SRF (specialised resource function). Depending on the capabilities of the particular implementation this could be internal to the SSP,[4] as discussed here, or externally located in an IN IP (intelligent peripheral) node. Procedures for connecting and communicating with IP nodes are described later in this chapter.

This same trigger point might be used to construct a 'hot line' service, whereby a dedicated line in a public location (such as an airport terminal or railway station) is connected directly to a pre-chosen destination – maybe a taxi company. In this case the SCP would return a 'Connect' message containing a destination number address for routing, instead of a 'Play Announcement' message, as in the previous example. Without IN, 'hot line' services are available, but usually at the expense of renting a private circuit from the network operator.

A further variation is the delayed off-hook trigger, where a call is automatically routed to a pre-arranged number if no digits are dialled within, say, 5 seconds of a phone going off-hook. This might typically be useful for an elderly person living alone, where the pre-arranged number could be that of a community help-line.

This variation of service type that can be achieved with the same trigger point illustrates the service independence that the IN infrastructure strives to achieve. The same basic function of an off-hook trigger has been used to construct several quite different services.

A third important trigger type, shown in Figure 2.4, is the terminating trigger, which applies to incoming calls to a particular subscriber. This could be used for a call forwarding service, as shown in Figure 2.4, where an incoming call is diverted unconditionally to another line. The trigger could equally be set to activate on 'busy' or 'no reply', so providing a conditional call forwarding service.

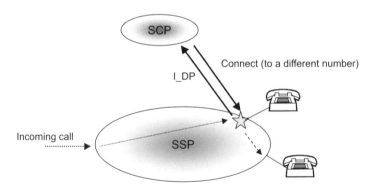

*Figure 2.4   A terminating trigger*

---

[4] **Note**: In an overview discussion such as this, the terms SSP and SSF are often interchangeable. Strictly, the SSF is the IN access **function**, whereas the SSP is the **point** at which that function (and possibly others) is located. Later on we are more careful when we need to distinguish between the physical SSP network element and the associated SSF software functional entity. The same goes for SCP and SCF.

## 2.2.2 Introduction to IN call models

State transition diagrams are normally used to chart the progress of a call through its various stages in call processing software, and a representative simplification of such an algorithm is shown in Figure 2.5.

Figure 2.5 shows an event sequence for a simple call, illustrating the call model concept and the consequent ISUP messaging. A caller lifts the handset, causing the state machine in the SSP to invoke an instance of the call model, starting it in the 'off-hook' state. When digits are dialled the call model goes to the 'receive digits' phase, then onto the 'analyse digits' state. When the call control software has decided where to route the call, the call model goes to the 'route call' state, and, assuming the call is destined for a user at a different exchange, an ISUP IAM (initial address message) is launched on an appropriate telephony route. When the far-end exchange has checked the address digits and has started to ring the destination user's phone, an ACM (address complete) message is returned, followed by ANS (answer message) if the called user answers. The call is now in the speech phase, and it is released when the users replace their handsets. This instance of the call model has now completed its job and can be terminated after any background housekeeping tasks, such as updating statistics and charging records, have been completed.

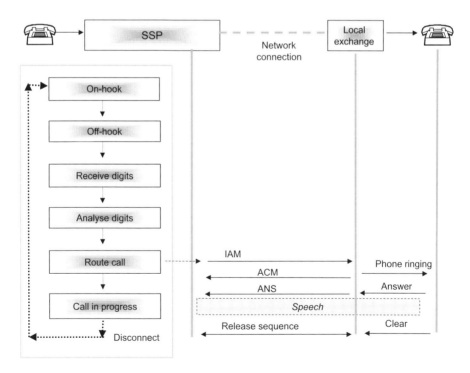

*Figure 2.5   A simple state transition diagram for a call set-up*

It must be emphasised that Figure 2.5 is a simplification, intended merely to illustrate the principles. It does not show all the states, nor does it use their correct names. The formal state model for ETSI Core INAP [10] is shown in Figure 2.8, where it will be seen that there are in fact simultaneous call model instances for the originating and terminating halves of the exchange call processing. In Figure 2.5 we only show the state diagram for the originating side; in other words we see the call from the caller's, not the called party's, viewpoint.

### 2.2.3 Trigger detection points

Triggering can occur at several places during the progress of a call, and Figure 2.6 shows four trigger detection points added to the simple call sequence shown in Figure 2.5. The larger rectangular boxes in the diagram represent stages of relative stability in the real-time call processing. They are, for this reason, termed 'points in call' (PICs) and they have been renamed here with the formal CS-1 modelling titles. The small boxes, the 'detection points' (DPs), are the points which generally indicate a transition between states, and are where the IN 'triggers' can be implanted. The full list of the ETSI Core INAP DPs is shown in Table 2.1.

IN triggers can be placed at several different points in the progress of a call set-up, but triggering need not automatically occur on every call that encounters these points. We can set various conditions that must be met before the processing is suspended and a request message launched to the remote service logic. If the conditions are not met then we can specify that normal call processing can continue, or be terminated, depending on the circumstances. Similarly, if the SCF is

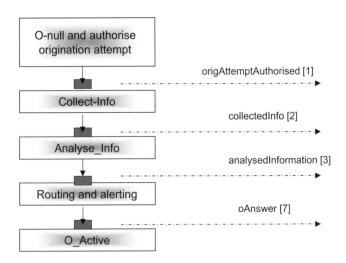

*Figure 2.6   Trigger detection points*

*Table 2.1   ETSI core INAP detection points*

| DP number | DP name |
|-----------|---------|
| 1 | OriginatingAttemptAuthorised |
| 2 | CollectedInfo |
| 3 | AnalysedInformation |
| 4 | RouteSelectFailure |
| 5 | OCalledPartyBusy |
| 6 | ONoAnswer |
| 7 | OAnswer |
| 8 | OMidCall |
| 9 | ODisconnect |
| 10 | OAbandon |
| *11* | *Reserved (not yet used)* |
| 12 | TerminatingAttemptAuthorised |
| 13 | TCalledPartyBusy |
| 14 | TNoAnswer |
| 15 | TAnswer |
| 16 | TMidCall |
| 17 | TDisconnect |
| 18 | TAbandon |

currently unreachable, we can fall back to normal processing, or apply a default routing, or fail the call after playing a suitable announcement to the caller.

### 2.2.4 Trigger types and trigger criteria

Most of the detection points listed above can be armed as Event Detection Points (E_DPs) as well as Trigger Detection Points (T_DPs – commonly known as 'triggers'). We discuss E_DPs later, and continue to discuss T_DPs here.

There are four categories of T_DP. These are:

☐ line-based (applicable to particular customers' lines)
☐ office-based (applicable across the entire exchange)
☐ private-facility-based (e.g. only applicable within a particular Centrex group)
☐ trunk-group-based (i.e. applicable to a single telephony route).

All of these, except office-based T_DPs, can be unconditional, meaning that they can be placed at many (but not all) DPs without any criteria being specified for them to trigger. Obviously this could be a dangerous privilege to give to office-based triggers – if every call passing through the exchange were to trigger, the recipient SCFs would be at risk of immediate overload, as would the exchange's call processing software itself.

Factors used as DP criteria include dialled digit strings, nature of address, CLI, class of service and call failure reasons. The set of relevant criteria obviously depends on the DP. For instance, a 'call failure reason' can only be expected at a DP which deals with call failures. The use of these as criteria is defined in (9), but in practice much depends on what is feasible in different implementations. There is scope for manufacturers to differentiate their products on different criteria selection availability, but much depends on performance and memory overheads imposed by an extensive range of criteria, as well as the complexity of software changes needed. It also of course depends on what the network operator can afford to specify, and this is likely to be based on service business case justifications for particular features.

### 2.2.5  Trigger tables

Trigger data is stored in SSPs in data tables. As well as the DP criteria data just discussed the SSP needs to hold information about what data to export when the SSF generates IN requests to an SCF.

#### 2.2.5.1  Trigger criteria data

Whilst the SSP does not contain any of the service logic for an IN service, it does need to be populated with data to enable the CCF to recognise at every DP whether the conditions and criteria exist that are necessary for a trigger to fire. This pre-set triggering data will have been set up as part of the IN service provisioning process, using the normal on-line data update mechanisms for exchange data.

#### 2.2.5.2  Trigger 'export' data

The SSP needs to send some basic information in the request message to the SCP for the service logic to perform correctly. The most important pieces of information are usually the addresses of the called and calling parties, although other information will be needed for different service circumstances. For instance the service logic might need to know the category of the line, in terms of whether it is a public phone box, or a business or residential line. This information may well have a bearing on the logic to be carried out for this particular service instance. For example, it may be inappropriate to offer a menu option of connection to a high-cost premium rate service, chargeable to the calling line, for a call attempt from a coin-box phone. If the trigger does occur in this case, the controlling service logic may be required to flag an error report to a log file or a console, and to instruct the exchange to connect dial tone and proceed as normal.

Similarly, the service logic may not be aware of the range of announcement capabilities in the requesting exchange and may need this information to be indicated in the request message. If the required announcement cannot be provided by the originating SSP, then the service logic may need to instruct the exchange to route the call to a separate network node – an IP, or maybe another SSP, which can deliver the correct announcement.

The set of parameters to be forwarded from a particular trigger situation will be defined in a trigger export-table in the SSP. To illustrate the data items that can be exported by an SSP in an IN message, Table 2.2 lists the 17 parameters specified in the ETSI Core INAP for the I_DP. As indicated in the table, only the first of these parameters is obligatory. Other parameters will be selected for inclusion depending on what information is needed by the SCF service logic to handle particular trigger requests.

An important feature of ETSI Core INAP is that an 'extension' mechanism is provided to enable extra fields to be included in any message. These are to cater for network-specific additions or for minor specification updates. To simplify the illustration the I_DP extension field is not included in Table 2.2.

### 2.2.5.3  I_DP parameters[5]

Just six of these 17 I_DP parameters were designed specifically for INAP. The remaining parameters have equivalents in the ISUP [11] or DSS1[6] [12] protocols, so if the underlying network signalling is ISUP then the parameters are simply copied across to the INAP message without change. We now describe these parameters individually.

**Service Key**. This parameter links the trigger event in the CCF to the relevant service logic program in the SCF. It enables the SCF 'front end' to route the request to the appropriate application SLP (service logic program) within the SCP. This

*Table 2.2   I_DP parameters*

| I_DP parameter | Parameter maps onto ISUP or DSS1 equivalent? | Optional (O) or mandatory (M) |
|---|---|---|
| serviceKey: | No | M |
| calledPartyNumber: | Yes | O |
| callingPartyNumber: | Yes | O |
| callingPartysCategory: | Yes | O |
| originalCalledPartyID: | Yes | O |
| locationNumber: | Yes | O |
| forwardCallIndicators: | Yes | O |
| bearerCapability: | Yes | O |
| eventTypeBCSM: | No | O |
| redirectingPartyID: | Yes | O |
| redirectionInformation: | Yes | O |
| iPAvailable: | No | O |
| iPSSPCapabilities: | No | O |
| cGEncountered: | No | O |
| additionalCallingPartyNumber: | Yes | O |
| serviceInteractionIndicators: | Yes | O |
| highLayerCompatibility | Yes | O |

---

[5] The parameter titles in Table 2.2 are presented using their ASN/1. (Abstract Syntax Notation No. 1) names. ASN/1 is the formal language used by the ITU-T and ETSI for defining application-level IN messaging.
[6] Digital Subscriber Signalling System Number 1.

linkage of a set of trigger data to a particular service key represents the extent of the CCF's 'knowledge' about the service details that are held in the SCF.

**Called Party Number**. For an originating local exchange trigger, this is normally the set of address digits dialled by the caller. In the case of a trunk exchange trigger it could be the contents of the called party number parameter received in an ISUP IAM on an incoming route. The purpose of the service logic could be to translate these digits into another routing address to be returned to the CCF in a consequent INAP message.

This parameter will obviously not be present in the case of an 'off-hook' trigger, because no dialled digits would be available at this stage of the call.

**Calling Party Number**. This identifies the call's origin, in terms of the originating user's network identity. In the case of a trunk trigger it would have been received in the signalling set-up message (e.g. the ISUP IAM as before).

**Calling Party's Category**. This indicates the type of the calling termination. Possible options include:

☐ operator (speaking a specified language)
☐ ordinary subscriber
☐ priority subscriber
☐ payphone.

**Original Called Party Identity**. This contains the digits the caller originally dialled. These may have been changed earlier in the call if a 'call forwarding' service had been encountered.

**Location Number**. This applies to calls from mobile subscribers. Because the Calling Party Number for a call from a mobile user gives no information about the caller's geographic whereabouts, the Location Number parameter is provided to carry data that can represent the geographical location of the call's originator.

**Forward Call Indicators**. This is a two-octet set of indicators containing information about how the call should be routed at intermediate exchanges. The indicators are as follows:

*National/international call indicator*. This tells the destination whether a call should be treated as an international or national call.

*End-to-end method indicator*. This specifies which, if any, method is to be used in the network for conveying end-to-end information. If this information is to be transferred, there are two options – the 'pass-along' method (using a dedicated ISUP message) or a connectionless network signalling procedure.

*Interworking indicator*. This says whether or not SS7 signalling has been used throughout the entire network connection for this call.

*End-to-end information indicator*. This indicates whether or not there is further end-to-end information available at the sending end. If there is, the receiving exchange will request it before alerting the called party.

*ISDN user part indicator*. This tells the recipient exchange that ISUP has been used all the way so far in the call connection set-up.

*ISUP preference indicator*. This says whether or not ISUP is preferred for the complete connection.

*ISDN access indicator*. This says whether the originating terminal is ISDN or not.

*SCCP method indicator*. This is for end-to-end transmission of information via SCCP, and says whether connectionless or connection-oriented (or both) methods are available.

**Bearer Capability**. This is a Q.931 information element that describes the connection type and the network signalling capability. It is carried in ISUP in a parameter called 'User Service Information' and provides values for items such as coding standard, information transfer rate and Layer 1 protocol type.

**Event Type BCSM**. This parameter is the detection point (DP) number for the trigger occurrence. Figure 2.8 shows the DP number on the BCSM.

**Redirection Party Identity**. This gives the network address from which the call was last redirected (diverted).

**Redirection Information**. This carries information such as redirecting reason, redirecting counter, etc.

**'IP Available' Indicator**. This says whether or not an IP is attached and available at the triggering exchange.

**IP SSP Capabilities**. This gives information on whether the triggering SSP has an internal SRF that could assist with the triggered service.

**Call Gapping Encountered**. This indicates if the triggered call has encountered (and survived) call gapping.

**Additional Calling Party Number**. This parameter is for the CLI provided by the calling user's access signalling system (rather than that provided by the serving local exchange).

**Service Interaction Indicators**. This parameter is intended for passing indicators about relevant switch-based services to the SCP. As the contents are not standardised they can vary greatly between implementations.

**High Layer Compatibility**. This gives information about the technical characteristics of the ISDN-Teleservice. Examples of teleservices follow (ITU-T recommendations I.240 and I.241) are:

☐ Group 4 fax
☐ Teletex (for exchanging text documents)
☐ Videotex
☐ 7kHz telephony (for high quality voice services).

### 2.2.6 Event detection points

So far we have been concerned with T_DPs, which are pre-armed trigger detection points, waiting to be 'fired' by call control software during normal call processing.

IN also has 'event detection points' (E_DPs), which are dynamically armed during a call set-up by instructions from SCFs. For example, the SLP may need to be advised if a call attempt meets a called party busy signal.

The INAP operation used by the SCF to arm an E_DP at an SSP is the RequestReportBCSMEvent. If the CCF subsequently detects this E_DP event, the SSF will respond to the SCF with an EventReportBCSM operation.

### 2.2.7 The CS-1 basic call state model

The purpose of the basic call state model (BCSM – defined in Q.1214 [9]) is to provide the SCF with a representative view of the call processing in the SSF. The SCF obviously has no need for the details of the call processing implementation. It just needs to know enough to enable it to process the service logic correctly.

The CS-1 BCSM postulates that call-control software in an exchange can be modelled as if a call has two halves – an originating half and terminating half (Figure 2.7). Two separate but interconnected call models, namely the O_BCSM and the T_BCSM, represent the call. The full Core INAP BCSM is reproduced in Figure 2.8.

Principally, what Figure 2.8 shows is that the originating (left-hand) side (O_BCSM) monitors the call at the incoming trunk or line, and the terminating (right-hand) side (the T_BCSM) monitors the outgoing side. The T_BCSM has fewer states than the O_BCSM. This is because the T_BCSM is not invoked until the O-BCSM is some way into the processing of a call, and it so exists for a shorter period of time.

In fact the majority of the common IN services can be constructed using only originating-side triggers. Hence, for an originating call from a local exchange, the O_BCSM handles the bulk of the set-up stages. It covers the initial authorisation (PIC 1), where the software checks that the caller who has just lifted his handset is in fact authorised to make calls and can indeed be supplied with dial tone. It then collects (PIC 2) and analyses (PIC 3) the digits the caller has dialled. It then launches the ISUP IAM (or equivalent) with the intention of alerting the called party (PIC 4).

PIC 4 is the stage at which the terminating half of the call model is invoked. This is indicated by the (dotted) communication line (1) in Figure 2.8. At the terminating side, the call enters the formidably named 'T_Null & Authorise Origination_

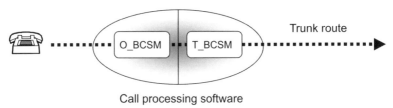

Call processing software

*Figure 2.7   '…. a call has 2 halves …'*

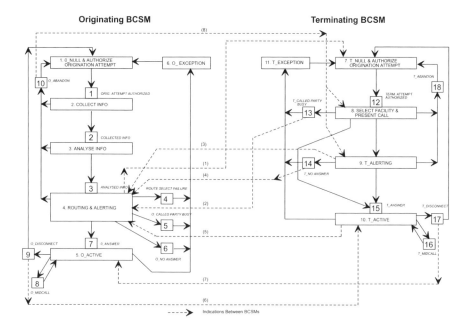

*Figure 2.8   The combined BCSM for CS-1*

Reproduced, with permission, from Figure 4–9/Q.1214 ITU-T publication 'Distributed Functional Plane for Intelligent Network CS-1' [10/95].

Attempt', moving through the subsequent states as the call progresses into the speech phase, where both sides of the call model achieve the stable 'active' state.

### 2.2.8 Notification and response DP types

It is possible to set a CS-1 detection point merely to inform the service logic that an event has occurred. This type of DP is called a T_DP-N (trigger detection point – notify), distinguished from a T_DP-R (trigger detection point – response), which is the one we have generally discussed so far. Call processing is suspended at a T_DP-R, awaiting a response, whereas CCF processing continues with a T_DP-N, and no reply is expected.

### 2.2.9 Triggering dynamics

The various triggering functions we have been describing are illustrated in Figure 2.9.

This diagram shows the interaction of the three main IN call processing functional entities, the CCF, SSF and SCF, for different T_DP types. It shows the basic call manager (BCM) functions in this role. The basic call manager

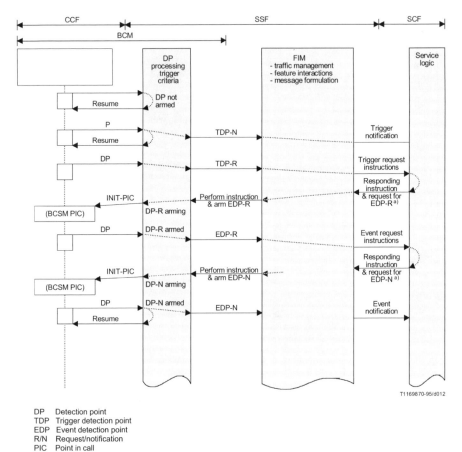

T116987 0-95/d012

DP    Detection point
TDP   Trigger detection point
EDP   Event detection point
R/N   Request/notification
PIC   Point in call

a) In this example, the responding instruction and request for EDP are shown together.
   These are independent information flows and may not be sent together in all cases.

*Figure 2.9   An illustration of the dynamics of triggering*

Reproduced, with permission, from Figure 4–10/Q.1214 ITU-T publication 'Distributed Functional Plane for Intelligent Network CS-1' [10/95].

encompasses the BCSM and the associated DP processing functions. Also introduced here is the CS-1 feature interaction manager (FIM).

Figure 2.9 shows five CCF trigger events occurring sequentially. The first is an unarmed DP (really a non-event), where a message notionally goes from the CCF to the SSF, where it is checked for arming criteria and found to be not armed. The next illustration is a T_DP-N trigger, where the call processing resumes immediately and the SCF is informed when the event occurs. This might, for instance, be used for a service that needs to log the number of calls sent to a particular destination address. Next is a T_DP-R, where call processing is suspended until a reply returns from the SCF. In this case the reply initiates a

BCSM instance in order to arm the required E_DP-R (event detection point – response), by sending an appropriate EventReportBCSM operation. When this DP is encountered, an E_DP-R response is sent to the SCF. Lastly, an E_DP-N (event detection point – notify) is returned to the SSF and when this event is encountered, the event is reported via the SSF, but call processing continues.

## 2.2.10 The CS-1 feature interaction manager

The FIM has the following main functions:

☐ **Traffic management.** This function checks to see whether any controls such as call gapping or service filtering are active before passing received messages to the SCF. Call gapping and service filtering are IN mechanisms (discussed later in this chapter and illustrated by examples in Chapter 7) that aim to reduce congestion problems at SCPs by reducing the quantity of messages sent from SSPs.

☐ **Feature interaction control.** Rules for controlling interacting features are applied here. The aim is to manage the current service request in the context of other outstanding service requests, or active service features, particularly switch-based service features that may have been invoked independently. For this reason, close co-operation with CCF feature management is needed here.

Also there are DP processing rules to be applied. For instance, a detection point could conceivably be armed both as a T_DP and an E_DP with the same trigger criteria, so consistency rules are needed to ensure that there is always a single point of control from the SSF. Obviously the 'executive' control point in the SSF should only communicate with a single SCF service-logic program-instance at a time in this event. Generally, Notification events and triggers take higher priority than Request events and triggers, and an E_DP takes higher priority than a T_DP if both are armed at the same DP. In this case, the T_DP is processed after the E_DP if (and when) the latter results in termination of the existing SCF control relationship.

A control relationship persists as long as there is at least one E_DP-R armed for at least one CS (call segment) of a CSA (call segment association). If only E_DP-Ns remain, the active control relationship changes to a monitor relationship, and it can coexist with a separate control relationship on the same call.

☐ **Message formulation.** This is the identification of the correct destination service control function and the formulation of the request message (e.g. an I_DP operation), with the required export data, to the appropriate signalling handling function.

## 2.3 Hosting and creating IN services

Whilst a major part of the cost of an IN implementation lies in the switch re-arrangements we have been discussing up till now, we now need to look at the functions outside the SSP that actually direct the service itself.

### 2.3.1 SCPs and SCFs

For early IN applications, in the late 1980s, the prime requirement was for an efficient external database for number translations. So the first SCPs tended to be service-specific, specifically designed for translating numbers such as those prefixed with digits such as 1800 (in the USA) for the universally popular Freephone services. In these implementations the transaction response from the SCP was a reply message containing data which was to be translated into a specific action by the SSP, which had knowledge of the specific context of the invoked service.

The term 'service control point' actually refers to a physical network node containing the hardware on which the control functions run, but it is often used to loosely represent the external service logic, and supporting structure, with which the SSP communicates when it activates a trigger request. The IN function providing the remote service logic is more properly termed the SCF (Service Control Function), and it is commonly located in a seperate SCP in traditional IN solutions. However it can be located in other places, including the SSP node itself. In this case, where the service logic is co-located with the call control function, the SCF is still 'remote' because it is functionally external to the normal call processing software.

The prime purpose of the SCF is to provide the software environment that allows the execution of the service logic programs whilst providing the necessary support functions such as signalling access and transaction control, logic program selection, provisioning and management. The SCF must be able to control interactions between, and simultaneous invocations of, multiple SLPs and SLP instances.

### 2.3.2 Service data considerations

Early IN services were provided on service control points in blocks of logic and associated data, which would include service housekeeping data and network routing data along with customer-specific data such as account details, service preferences and redirection numbers. A danger here is that when customers subscribe to several services their identity data (CLI, VPN groups, etc.) might be provided several times over. This can come about if service 'products' are marketed and implemented separately within a sales department of an operator's administration.

A more flexible approach can be to store 'profiles' of customer data, so that when a customer is involved in any service invocation, that customer's particular profile is accessed, rather than a block of service-specific data. Using this scheme, data are organised into customer-specific, rather than service-specific, data. These 'user profiles' are then available for all IN service invocations. The focus switches from the service to the customer, who benefits from a unified set of services. This is easier to use than a set of discrete service packages, with perhaps different authentication passwords for different service applications.

ITU-T defines the service data function (SDF) for the management of IN data, and identifies an SCF–SDF interface for accessing the data. The ETSI Core INAP standard SCF–SDF interface is defined in [13]. The SDF can be located in an SCP (or other) node, with the associated SCF service logic, or in a dedicated service data point (SDP), where the data is to be accessed from several different SCPs in the network.

The SCF–SDF interface is sometimes used to access switch-based data from an SCP. An example would be where an SCF service needs to make a decision based on the state of a switch-based call diversion service. The service logic may require the SCF to retrieve the address digits of the customer's diversion number. In this configuration the SSP can be viewed as including an SDF, and the SCF–SDF INAP interface is in parallel with any existing SSF–SCF interface.

### 2.3.3 Multi-service SCPs

The growing interest in virtual private networks led to the use of specific instructions from the SCP to the SSP. This started the evolution of the SCP from a relatively simple database facility into a wider network control point, which housed service logic as well as the database. This trend continued towards what was later realised to be unrealistic limits with the Bellcore proposals for IN/2, which are further discussed in Chapter 4. IN/2 attempted to move all service logic to the SCP, ignoring existing switch implementations that also contain supplementary service logic.

A typical architectural structure for a multi-service SCP is shown in Figure 2.10. This shows the SLEE, which acts as the 'front-end' of the SCP, passing requests and responses onto the correct ('back-end') service logic program (SLP). The SLP is referenced by the 'service key' (SK), which is inserted into I_DP messages by the SSF (see Section 2.2.5). The SLEE is discussed in more detail below.

### 2.3.4 Service logic execution environment (SLEE)

Externally, the SLEE appears as the 'front-end' for the SCP, because, as already discussed, it arranges for incoming requests to be handled by the appropriate service logic program. It actually does far more than this, as its prime role is to host the SLPs. The SLEE provides the operating system for the execution of the SLPs, and presents everything that is needed in a platform for service execution to its user applications.

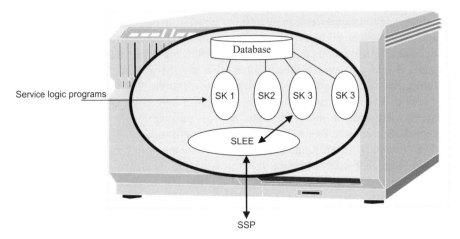

*Figure 2.10   A multi-service SCP*

The SLEE provides management and housekeeping interfaces for SLPs, enabling their associated objects to be instantiated, activated, deactivated and withdrawn. It provides and manages local resources needed by the SLPs, such as transient call data, threads, clocks and synchronisation utilities. The SLEE will co-ordinate service execution, providing scheduling and prioritising services; it will monitor for errors and issue fault reports where needed.

The SLEE also organises the database services needed by the SLPs for their service data, which is provided by an SDF. The SDF can be co-located with the SCF at an SCP, as indicated in Figure 2.10, and data can also be requested from an external SDF, hosted in a dedicated SDP. This would be the case where, for instance, customer profile data have shared access from SCFs at several SCPs.

### 2.3.5 Service creation environment

Chapter 1 discussed the original 'jewel in the crown' for the IN vision, which was the prospect for network operators to be able to control the creation and modification of services, and to be able to introduce them to their real-time networks quickly. This would do away with the traditional long and expensive process of ordering and negotiating new services from equipment suppliers and enable operators to corner niche market opportunities before competing operators. The term 'service creation environment' (SCE) refers to any platform designed to meet this need for crafting new services that can be hosted on an IN SCF.

The standard IN architectures (such as Figure 1.5) show the SCE function (the SCEF) with an interface to a service management function (SMF), which, in turn, interfaces to the SCF. The SMF has a secure interface to the SCF, and so SLPs that have been created and tested at the SCEF are downloaded to the SCF under carefully controlled conditions by the SMF after the SMF is satisfied that the SLP is network ready.

Traditional telephony services undergo rigourous testing before being brought into service. In the SCE approach to service development therefore it is necessary to provide tools for off-line checking and simulation before the service is allowed to operate in the live network. As well as SSP and SCP simulators, network operators will typically have captive installations for pre-service testing, and operational SCFs are likely to have facilities for enabling the service to be trialled in a controlled environment before full launch.

From a user's viewpoint, the SCE function usually takes the form of a graphical interface on a computer screen. This enables the designer to build up a representation of the service logic pictorially, using a set of defined building blocks and procedures. The service building blocks are called SIBs and they are discussed further in Chapter 4.

A common technique in service creation systems is the use of a 'tree' structure for expressing the flow of the service logic. An example of a segment of a simple incoming call management service, where calls to a customer's number are diverted to a voice mail system at certain times, is shown in Figure 2.11.

### 2.3.6 Operational support systems (OSSs) for IN services

Intelligent networking is really about the fast introduction and efficient delivery and management of a multitude of services, and this depends as much on the operational 'surround' as on the complexities of the service logic in the SCP and the associated real-time IN network components.

There are two broad categorisations for OSS, and these are service management and network management – although the two are of course strongly inter-related.

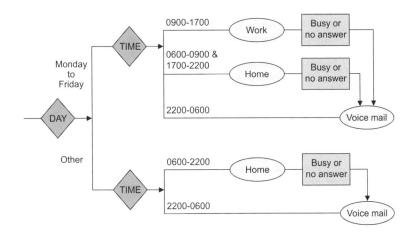

*Figure 2.11   A service creation tree*

We return to service and network management several times during the course of this book, but for the time being some of the main functions provided by the two disciplines are illustrated in Figure 2.12.

Service management has overall control of the initial loading of a new service and its data, rollout of new services, taking orders for the services from potential customers, organisation of billing and revenue-apportionment schemes and, generally, all customer and business-facing issues.

Network management is concerned with the smooth operation of the network itself and it has responsibilities associated with network planning, configuration and upgrade (to provide the infrastructure required for the IN services). It also has responsibility for network data updates – e.g. for SCP number translations, SSP trigger tables[7] and network routing data – and for in-service support issues such as maintenance, software and hardware routing, fault diagnosis and fault clearance. Network management is also responsible for the performance aspects of network control, such as statistics production and traffic congestion control. Last, and certainly not least, the successful transfer of call charging records from their points of origination to the appropriate billing centres is also a network management responsibility.

As we imply in Figure 2.12, the service management function pervades most IN elements, although, as we indicate, it is likely to have lesser involvement with the IN IP because this does not generally hold volatile service-related data. However, an efficient service management data update route is needed for service data held at

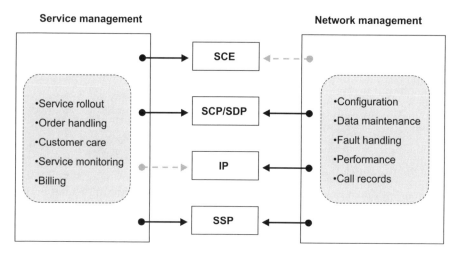

*Figure 2.12   Management environment for IN network elements*

---

[7] Section 2.2.5 describes the SSP IN trigger data and Section 2.6.3 further discusses some particular issues relating to SSP data management complexities.

the SCP and SDP nodes, and, for changes to service triggering parameters, the service management needs to have access to trigger table data held on the SSP.

Network management functions are mainly targeted on the high performance real-time elements. These are primarily the SSP, IP, SCP and, to a certain extent, the SDP (for the case where a service's data is stored separately from the service logic in the SCP and needs to be accessed in real-time). However, the SCE is likely to have less involvement with the network management systems, so a reduced linkage is shown between SCE and the network management in Figure 2.12.

Generally speaking, standards are available for management interfaces in the lower part of the diagram (for the real-time elements) [49–51]. However, the applicability of standards decreases as we ascend through Figure 2.12. This is because, as we move upwards into the SCE area, the variety of the proprietary interfaces in existence tends to increase because we are approaching the domain of business-specific and necessarily non-standardised methods and policies for service strategy, revenue generation and customer relationships.

In most cases the design of the service logic and the real-time IN architecture is actually a smaller task that that of organising the support 'surround' needed for the new service to work in the operational environment suggested by Figure 2.12. OSS covers an extremely wide range of network support activities, as suggested by the example lists of functions in the diagram. The details of operational support systems (OSSs) are outside the scope of this book, and the reader is referred to [48, 49] for coverage of service and network management systems and their standards. Chapter 4 below does include some further discussion on the service management recommendations that were published for the ITU-T IN CS-2.

### 2.3.7 SCP 'non-functional' requirements

We have so far described the 'functional' requirement of SCPs, i.e. the things they are expected to do to perform their main task of responding to requests for assistance from the underlying switched network. However, there are several 'non-functional' expectations for SCPs; these are necessary supporting attributes, or behaviour characteristics.

#### 2.3.7.1 Continuous availability

Telecommunication systems are traditionally renowned for their rigourously high standards of availability – a factor that has always fundamentally influenced the design of central processor units in digital exchanges. Transaction processors used by banking systems, for instance, must have high standards of security in terms of data integrity, and must have comprehensive methods of roll-back and roll-forward to ensure that data are preserved to the last detail under failure conditions. However, even these systems are allowed 'down-time' at quiet periods, for upgrades and maintenance activities. In the case of telecommunications equipment, even a 5 minute planned outage in the middle of the quietest night is often regarded as intolerable. A distinctive feature therefore of the design of telecoms control equipment is the provision of mechanisms which allow real-time on-line data updates. These mechanisms allow data updates to be made whilst the system is operating, usually at normal load. They also allow software enhancements to be carried out while the system is operating. In practice, a system 'restart' period of perhaps 30 seconds (usually at night) is often tolerated, to allow the changeover to the new software, but this is usually conditional on there being no calls in progress lost.

Unsurprisingly, therefore, performance criteria for SCP nodes are usually equally rigid. However, for SCPs, this can be achieved by redundancy, using alternative network routing arrangements, multiple sites, diverse routes, redundant communication links, or a workable combination of all of these. In the early years of intelligent networking, SCPs were based on digital exchange designs, because of their proven availability standards. Following this, there was a move towards banking-community transaction processing machines. However, because of their extremely high-standard security requirements, transaction processors are expensive. It was gradually realised that data in telecommunications does not perhaps have quite such high security needs, and that although the 'fault-tolerant' transaction processors are often advertised as 'non-stop', they were not continuously available – requiring down-time for system maintenance and upgrades. Nowadays the trend is towards achieving full availability through over-provisioning of cheaper, less fault-tolerant but high performance computing units.

However, the critical issue for continuous availability is software reliability, particularly in complex software systems. The current trend is towards object-oriented design, because experience with systems which have been developed using these techniques has generally demonstrated higher productivity and lower fault rates.

### 2.3.7.2 High performance

The aim is to optimise the system's performance, particularly at the high end of its design load. Overload protection mechanisms should not be allowed to compromise system performance. On the other hand, an application's performance achievement should not suffer because a neighbouring service application receives a large amount of transaction requests. Flow control mechanisms should be sufficient to protect the non-offending application process. There is often a need for resource reservation schemes to preserve processing capacity for high revenue-earning services, or for particularly important customers. The provision of a single resource pool for the whole node is usually more efficient than providing local 'pockets' of such functionality.

### 2.3.8 Future trends for SCPs

So far we have mainly described IN architectures with centralised service logic, implicitly hosted on stand-alone, monolithic, SCP processors. The configurations discussed have, for simplicity, shown SCPs connected to many SSPs via SS7 networks using relay points and signal transfer points. Data and programs would be loaded via dedicated service management systems, directly connected to the SCPs, which are often configured in pairs for security. Data would be mastered on a separate database, or one of the SCP's internal databases would act as 'master' and the other SCP's database would follow in step with data changes.

However, SCP evolution trends have generally tended towards more flexible computing structures with arrays of co-operating servers hosted on a bus and rings. This introduces the concept of multi-processing to SCPs, providing easy scalability, enabling the addition of new functionality servers when required. The design trend has continued towards the prospect of fully distributed computing systems, when the ring structures have expanded and effectively become wide-area networks connecting geographically dispersed servers and databases. Distributed service logic is discussed further in Chapter 6.

## 2.4  The intelligent peripheral

Much of the discussion so far in this chapter has centred on the relationship between the SSF and SCF and how we can reach out from normal call processing to supplementary service logic which can change the course of the call, depending on data held there. We saw in Chapter 1 that many IN voice services also need some interaction with the caller, either to play an announcement or to collect some further information to supplement the dialled digits to allow the service logic to proceed to completion. For example, a caller may be offered a choice between long-distance carriers, so a menu of options would be presented verbally and callers invited to indicate choices verbally, or by pressing key-pad digits.

To provide this facility the IN IP was introduced in order to temporarily terminate a speech channel on behalf of the IN service logic and to provide a means of collecting further instructions from the user about the subsequent call handling. The SSP will already have a repertoire of announcements, such as those that can be played following receipt of call rejection messages in normal call handling. It may be that the IN service logic needs to instruct the SSP to play one of these standard messages, or it may be that a specialist message is required. In the latter case the call could be routed to an IN IP.

POTS users have a limited signalling repertoire available. Traditional dials used make-and-break pulses, which can only communicate with the network using the digits 0 to 9. The more common DTMF key-pads are slightly better off, having * and # keys so that codes for activating switch-based services such as call diversion can be used. In the early days it was simply not envisaged that users would want to use their signalling interfaces to do anything more that dial destination telephone numbers comprising digits between 0 and 9.

Standards are not specific about whether the IP should be seen as a network node or terminal equipment, and both options are included in the ITU recommendations. If it is to be a network node, the natural interface to the SSPs is ISUP, which is a network signalling system; if the IP is to be terminal equipment, a customer signalling system (such as DSS1) is appropriate, and is more convenient for terminal manufacturers. The question hinges on whether the IP is seen as a public or private resource, and there are identified needs for both.

Whilst we have introduced the IP as a new network node for these user-interaction features, the functionality needed (termed the 'specialised resource function' – SRF) could physically be located in an SSP. In fact SSPs almost always have a degree of SRF built into them, e.g. for playing general-purpose announcements. However, for more specialist dialogues the SRF can be located in a separate IP node, as illustrated in Figure 2.13.

The SRF acts on behalf of the SCF. It is the SCF that arranges for an SRF to be brought into a call and the SRF in turn reports its results back to the controlling SCF. When the SCF service logic decides that it needs further information from the caller in order to proceed, it instructs the SSF to arrange for the caller to be connected to an appropriate SRF, which in Figure 2.13 is located in an external IP.

There are several variations on the mechanisms used to bringing an IP into play for a call. One of the commonly used sequences is known as the 'Assist' procedure, which is described in the next section. The Core INAP operations available for the SCF to instruct the SSF to make a connection to an SRF are the 'establish temporary connection' (ETC) and the 'connect to resource' (CTR) operations. The CTR is typically used for connections to SSP internal functions, and the ETC (illustrated in Section 2.4.1) is generally used for connections to external IPs.

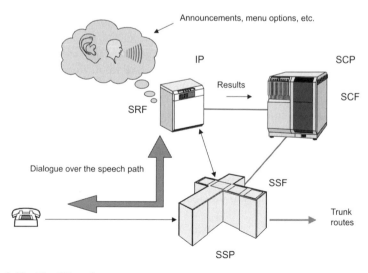

*Figure 2.13   The IP's role*

### 2.4.1 The Assist procedure for IP connections

Figure 2.14 shows the network messaging relationships[8] and Figure 2.15 shows the sequence of messages between the elements. We assume that the IN service logic is invoked by an I_DP message from an SSP (message 0 in the diagrams).

After the appropriate SCP has been invoked by message 0, the SCP instructs the SSP to establish a temporary connection with a nominated IP node (message 1), and the SSP sets up a bearer connection (message 2). The ETC message contains two essential co-ordinating parameters that must consequently be passed forward to the SRF in the IP node in the bearer signalling. These two parameters are the 'SCF identity', which enables the IP to identify the controlling SCP and the 'correlation ID', which is a reference number sent directly from the IP to the SCP in the ARI message (number 3). This enables the SCF to associate this new transaction opened with the IP with the original transaction with the SSP.

Having made this association, the SCF sends a 'prompt and collect' message (message 4) to the SRF instructing the speech resource equipment to present audible choices to the caller. It also instructs the SRF to prepare to receive a response from the caller over the transmission path and to return this to the SCF. Because there will be a delay whilst the user–IP interaction takes place, the SCP instructs the SSP (message 5) to reset its covering timer for a response to the original I_DP.

After information has been obtained from the caller by the SRF, it returns (message 6) a 'return results' message in reply to the 'prompt and collect' of message number 4. The SCP hopefully now has enough information to instruct the SSP on what to do next, such as route the call to a particular destination address. The SCP therefore ends the dialogue with the IP (message 7) and instructs

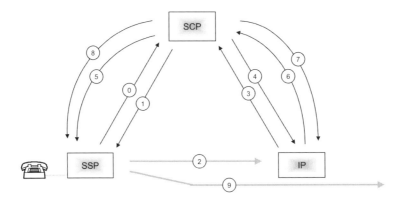

*Figure 2.14   Assist procedure messaging relationships*

---

[8] Acknowledgment is made here to the ITU-T recommendation for the Assist procedure description in Q.1219 Appendix 4.2.

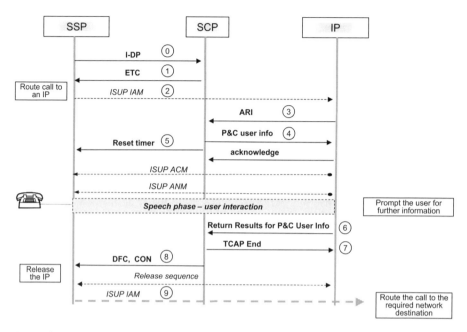

*Figure 2.15   The Assist procedure – message flow*

(message 8) the SSP to disconnect the 'call' to the IP and to set up a call attempt to the required destination (message 9).

The bearer signalling transport protocol between SSP and IP depends on individual network arrangements. Commonly used transport protocols are ISUP (used in our example), TUP derivatives and Q.931. ISUP has the co-ordinating parameters specified and they can be carried in the initial address message (IAM) but this is not the case with TUP derivatives, such as IUP. In this case the parameter contents might be included as a prefix to the address digits in the IAM (or IFAM).[9] It will then of course be necessary to work around any local limitations caused by number length restrictions.

In summary, the numbered stages in Figure 2.14 correspond to those in the message flow of Figure 2.15, representing the following actions:

0. I_DP from SSP to SCP.
1. ETC from SCP to SSP.
2. The SSP routes the call to the IP.
3. IP sends ARI (Assist request instructions) to open a transaction with the SCP.
4. SCP sends P&C (prompt and collect) back to the IP.

---

[9] This may also be the case with some implementations of ISUP. Although the field specification for SCF Id. and Correlation Id. are included in Q.763 [7] corresponding procedures were not specified in earlier standards.

5. The SCP instructs the SSP to reset its application timer.
6. The IP returns the collected information to the SCP.
7. The SCP terminates the transaction with the IP.
8. The SCP sends DFC (disconnect forward connection) to the SSP to disconnect the temporary connection to the IP, and Connect is also sent to instruct the SSP to route the call to the required network destination.
9. The call is onward routed.

The use of the 'Assist' procedure described here is further illustrated in Chapter 7, where we show how it can be used several times in the same call to offer a 'call completion' facility for a directory enquiry service.

## 2.5 INAP (the intelligent network application protocol)

The real-time IN architecture elements (SSP, IP and SCP) and the associated functions (SSF, SRF and SCF) have now been introduced and so we next return to the INAP messaging system that links these IN components together.

### 2.5.1 ETSI Core INAP operations

We provide here brief accounts of the 29 CS-1 INAP operations that were chosen for inclusion in the ETSI Core INAP specification. Complete descriptions of the procedures associated with these operations can be found in Section 3 of [10]. There are 26 ITU-T CS-1 operations that were **not** included in ETSI Core INAP. For background information these are listed separately in Appendix 2. The ETSI Core INAP operations are listed in Table 2.3 and then explained in the subsequent text.

**ActivateServiceFiltering (SCF → SSF)**

This operation was designed particularly for mass-calling applications, such as telephone voting ('televote'), where there are likely to be many similar requests to the SCF in a short space of time, all of which need to be counted. For more efficient traffic handling it is possible for the SCF to prescribe the standard action to be taken within the SSP, rather than trigger to the SCF for each call that meets the triggering criteria for this service. The procedure effectively devolves much of the repetitive part of the service handling to the SSF by allowing the SCF to instruct the SSF to only trigger for this service on a proportion of calls. The triggering might occur perhaps on every 20th or every 1000th call, depending on the level of demand expected for the particular voting scenario.

The calls to be filtered can be identified by using the triggering service key, or a combination of the triggering service key and the dialled digits. The criteria can be further narrowed to a particular geographic area by specifying some of the calling address digits. The action prescribed by the SCF could be typically to increment a counter and play a specified announcement to the caller. The values of the counters

*Table 2.3    ETSI Core INAP operations*

| 29 ETSI Core IN operations | SSF — SCF direction | SRF — SCF direction |
|---|---|---|
| ActivateServiceFiltering | ← | |
| ActivityTest | ← | |
| ApplyCharging | ← | |
| ApplyChargingReport | → | |
| AssistRequestInstructions | → | → |
| CallGap | ← | |
| CallInformationReport | → | |
| CallInformationRequest | ← | |
| Cancel | ← | ← |
| CollectInformation | ← | |
| Connect | ← | |
| ConnectToResource | ← | |
| Continue | ← | |
| DisconnectForwardConnection | ← | |
| EstablishTemporaryConnection | ← | |
| EventNotificationCharging | → | |
| EventReportBCSM | → | |
| FurnishChargingInformation | ← | |
| InitialDP | → | |
| InitiateCallAttempt | ← | |
| PlayAnnouncement | | ← |
| PromptAndCollectUserInformation | | ← |
| ReleaseCall | ← | |
| RequestNotificationChargingEvent | ← | |
| RequestReportBCSMEvent | ← | |
| ResetTimer | ← | |
| SendChargingInformation | ← | |
| ServiceFilteringResponse | → | |
| SpecialisedResourceReport | | → |

are then periodically returned to the SCP using the ServiceFilteringResponse INAP operation (explained below). Further illustration of the use of the ASF (ActivateServiceFiltering) feature is given in a televoting service example that is included in Chapter 7.

**ActivityTest (SCF → SSF)**

The ActivityTest operation is used to check that a previously established relationship between the SCF and SSF is still in existence. If it is, then the SSF will respond. Otherwise the SCF deduces that the relationship has failed in some way and it will close the outstanding transaction. This procedure might be used, for example, where an SCF had armed an E_DP for 'no reply' (at detection points 6 or 14 on the BCSM, Figure 2.8) and the call is a long duration one. As a safeguard the service logic may check that an EventReportBCSM response has not been lost by sending an ActivityTest operation to the SSF.

**ApplyCharging (SCF → SSF)**

The ApplyCharging operation provides a way for the SCF to interact with switch-based charging mechanisms.

**ApplyChargingReport (SSF → SCF)**

This operation is for the SSF to report back to the SCF in response to a previously received ApplyCharging operation.

**AssistRequestInstructions (SSF → SCF, or SRF → SCF)**

The ARI operation is used after a connection has been set up to an IP (or another SSP) as a result of a ConnectToResource or an EstablishTemporaryConnection operation being sent from the SCF to the original SCP. The assisting SRF, or SSF, will contact the SCF with the ARI operation in order to obtain information on how to proceed with the call.

**CallGap (SCF → SSF)**

The SCF uses this operation to tell the SSF to reduce the rate of I_DPs for a specified service key or calling/called address (or both).

**CallInformationReport (SSF → SCF)**

The CallInformationReport message is used as a reply to a previous CallInformationRequest operation.

**CallInformationRequest (SCF → SSF)**

This operation is used by the SCF to instruct the SSF to keep a record of specified information (about a single call) and respond, with the results, to the SCF at the end of the call using a CallInformationReportoperation. The requested information could be any of the following:

- ☐ call set-up elapsed time
- ☐ call stop time
- ☐ call connection elapsed time
- ☐ called address (before translation)
- ☐ call release reason.

**Cancel (SCF → SSF or SCF → SRF))**

This operation can be used to either cancel all outstanding requests or to particularly cancel a previous PlayAnnouncement or a PromptAndCollectUser-Information operation.

**CollectInformation (SCF → SSF)**

The CollectInformation operation is intended for use in conjunction with a RequestReportBCSMEvent operation in the arming of DP2. It specifies how many digits should be collected.

**Connect (SCF → SSF)**

The SCF sends a Connect operation to instruct the SSP to route the call to a specified destination. It is one of the most commonly used INAP operations and it enables the SCF to exert considerable influence over the CCF call processing. It can have the following parameters:

- □ **destination routing address**
- □ **correlation identity**. This is used for the 'hand-off' or 'Assist' procedures, and the intention is that the SSP should pass it forward (with the SCF identity indicator) in the network signalling system to the next SSP or IP node, which should in turn pass it up to the SCF. This will enable the SCF to correlate the new request with the original transaction with the first SSP.
- □ **SCF identity** (ditto).
- □ **cut and paste**. This allows the SSP to replace some of the leading digits of the address dialled by the caller with translated digits returned from the SCF.
- □ **route list**. This allows the SCF to over-ride the CCF's normal routing function by specifying a list of the telephony routes to be attempted to reach the call's destination.
- □ **calling party's category**. This allows the SCF to change the value (operator, payphone or ordinary customer) of this parameter from what it might have been previously set to, either in the exchange data for the customer, or as received in an incoming network signalling message.
- □ **original called party identity**. This is for the original number dialled by the caller.
- □ **redirecting party identity**. This is the address from which the call was last redirected en route to the SSP.
- □ **redirection information**. This contains information on the call forwarding history of the call, such as redirecting counters.
- □ **alerting pattern**. This is applicable for terminating local exchanges only, and allows different ringing patterns to be specified – assuming the exchange supports them.
- □ **service interaction indicators**. This is intended to provide a means of controlling interactions between switch-based network services and services provided at the SCF. The content, however, is largely unspecified, and so is highly implementation-dependent.

**ConnectToResource (SCF → SSF)**

The ConnectToResource (CTR) operation is used to instruct the SSP to connect a call to the physical entity (e.g. an external intelligent peripheral) containing an SRF. If this physical entity is internal to the SSP, the SCF should provide the call leg identifier for the leg to be connected to the resource (the A leg is the likely default) as a parameter. If the SRF is in an external IP, then an IP routing address will be needed.

**Continue (SCF → SSF)**

'Continue' is used by the SCF to request the SSF to proceed with call processing from the point at which it stopped in order to send the request to the SCF.

**DisconnectForwardConnection (SCF → SSF)**

This operation is used to disconnect a temporary connection, which had been set up (using EstablishTemporaryConnection or ConnectToResource operations), to a separate network resource such as an IP.

**EstablishTemporaryConnection (SCF → SSF)**

This operation, ETC, establishes a connection to a separate network resource, such as an IP, in order to play an announcement or to collect user information.

**EventNotificationCharging (SSF → SCF)**

This operation is used to notify the SCF of a charging-related event previously requested by the SCF in a RequestNotificationChargingEvent operation.

**EventReportBCSM (SSF → SCF)**

This is used to notify the SCF of the occurrence of a call-related event (such as 'busy' or 'no reply') that had been armed using a RequestReportBCSMEvent operation.

**FurnishChargingInformation (SCF → SSF)**

This operation can be used to send charging information to the SSF for inclusion in the switch-based call record to be sent for off-line processing at the end of the call.

**InitialDP (SSF → SCF)**

This is the first operation in an INAP interchange instigated by the SSF. It is sent when an active trigger has been encountered at a T_DP when all triggering conditions have been met and the call is not gapped or service-filtered. The 'service key' parameter is mandatory and there are 16 optional parameters (e.g. calling and called numbers); the available parameters are further described in Section 2.2.5.3.

**InitiateCallAttempt (SCF → SSF)**

This operation is used to request the SSF to create a new call leg, using address information provided by the SCF. It could be used to connect a destination number to an announcement device for an alarm call service.

**PlayAnnouncement (SCF → SRF)**

The PlayAnnouncement operation instructs the SRF to play a specified announcement, or tone, over the speech path. The speech path is set up using ETC or CTR operations. For standard tones and announcements the SRF can be collocated with the SSF in the network switch.

**PromptAndCollectUserInformation (SCF → SRF)**

This operation, P&C, is used to enable the SCF to interact with the user by instructing an SRF to prompt for and collect information such as in-band dialled digits, or voice responses, over the speech path.

**ReleaseCall (SCF → SSF)**

This operation requests the SSP to release an existing call, which can be at any phase of the set-up. A 'cause' parameter is provided.

**RequestNotificationChargingEvent (SCF → SSF)**

This operation is used to request the SSP to monitor for a charging-related event, then to respond to the SCF when the event occurs.

**RequestReportBCSMEvent (SCF → SSF)**

This operation is used to arm E_DPs, thereby instructing the SSF to arrange for the CCF to monitor for the occurrence of call-related events (such as destination busy or no reply). Several E_DPs can be armed simultaneously, but individual EventReportBCSM operations must be returned when events are detected.

**ResetTimer (SCF → SSF)**

This operation is used when SCF processing is taking longer than expected, and the SCF sends ResetTimer in order to instruct the SSF to refresh its application timer to give the SCF sufficient time to finish the processing for this request.

**SendChargingInformation (SCF → SSF)**

This operation requests the SSP to modify the real-time charging with information passed from the SCF. Because charging mechanisms vary significantly between vendors, the parameter content detail is operator-specific.

**ServiceFilteringResponse (SSF → SCF)**

This operation is used to reply to the SCF with the values of counters specified in a previous ActivateServiceFiltering operation. An example of the use of SFR (ServiceFilteringResponse) is given in Chapter 7.

**SpecialisedResourceReport (SRF → SCF)**

This operation is the reply to the PlayAnnouncement operation when the 'announcement completed report' indicator is set.

## 2.6  IN CS-1 implementation issues

As we have discussed, every exchange design is different, and so the ease of upgrading to IN varies between exchange types, and the consequent effect on overall performance overhead is important. The addition of triggers and the trigger criteria is a call processing upgrade in the switch in just the same way as any other call processing upgrade. A new switch build containing IN features is therefore subject to normal upgrade procedures, and IN features are likely to be competing against business-as-usual items like bug fixes, maintenance improvements and new (non-IN) service features. There are usually too many candidate functions for a single new build to handle and so the choice of what features go in a new build is inevitably a question of business priorities.

Because of these issues to do with upgrading existing switches, IN was typically introduced in a piecemeal fashion. The international standards allowed operators and equipment vendors to negotiate controlled sub-sets of the ETSI CS-1 functions, and the contents of these IN 'phases' depended on short-term requirements and budget allowances.

Because many useful IN services could be provided with just a limited set of SSP functionality, an early delivery of IN functions might have included, say, 5 of the 'originating' side trigger points in the BCSM and perhaps 12 or so of the most useful INAP operations. Obviously service 'walkthroughs' would be needed to make sure that the planned services would work with the chosen subset. Such a subset would have been designed to support the features needed for the more popular IN services involving number translation variants. More detailed terminating functions needed for, say, call queuing and call distribution controls could then be added in future builds in the light of ongoing service and business demand and budget availability.

### 2.6.1 Core INAP as a compliance benchmark

Discussions in this chapter are based on the CS-1 ETSI Core INAP, because, within Europe, this still remains the most widely used basis for IN implementations. The ETSI standard was itself derived from the ITU-T CS-1 standard, which in turn provides a wider framework, enabling linkage with world-wide trends in IN, in particular with US AIN standards.

The CS-1 BCSM was chosen by the ITU-T IN standards community as representing a generally acceptable common starting point for IN design. However, in practice, software designers of the industry's different digital exchange variants had previously developed their own pragmatic design models that would meet their particular digital exchange requirements.

The BCSM is a theoretical model, devised independently and after the designs of the digital exchanges themselves. Several modelling philosophies were possible, and so the choice became a theoretical compromise, but the BCSM was finally agreed as a suitable starting point for the standards production. The main purpose of the standards process was to equip the telecommunications industry with a usable basis for IN designs with as much international commonality as possible. The main goal was to develop an IN signalling protocol that handled the separation of the service logic from the legacy call control software in the exchanges. SS7 INAP filled this role, and the BCSM provided the necessary level of agreement across the international standards community.

It must be recognised that, from the SSP viewpoint, the IN standard in many ways still remains at the level of a useful theoretical construct, and cannot really be taken as a blueprint for the way things actually work in detail. Different switch manufacturers adopted the standard and amalgamate it with their particular call processing designs in their own ways. This then presents difficulties for operators when trying to assess which is the best product option for their network upgrade to IN.

At the SCP these concerns do not of course apply. Generally the BCSM is sufficiently detailed for the new SCF designs. The BCSM provides a 'birds eye' view of the essential switch functions, with the purpose of enabling the remote service logic in the SCF to be aware of the context of IN request transactions. The SCF obviously has no need for the details of the call processing implementation; it just needs to know enough to process the service logic correctly.

ETSI Core INAP [10] provided network operators with a practical text against which to check compliance of the various vendors' CS-1 IN implementations. For instance, ETSI CS-1 specifies 17 DPs (most of which can be T_DPs or E_DPs) and 29 INAP operations. Whilst it might possibly be feasible for an individual implementation to justify including all the 29 INAP operations, the justification is unlikely to extend to a full set of all the possible T_DPs and E_DPs. Therefore the standard has an invaluable subsidiary role as a compliance benchmark against which to check different manufacturers' available product sets.

*2.6.2  INAP implementations and feature interactions*

Within a chosen set of INAP operations there are still decisions to be made about which parameters are needed in the controlled sub-set. For example, the ETSI INAP Connect operation has 12 possible parameters, and these are detailed in Section 2.5.1 above, but some of these may not be relevant to some switch designs or may be expensive to implement. For instance, the parameter 'alerting pattern' specifies different ringing patterns for different calls, but how far this can be used of course depends entirely on the underlying switch implementation. Whilst the inclusion of a certain parameter in a message might be straightforward, providing the underlying supporting functions to make proper use of this parameter may be less straightforward.

The workings of many parameters are left for the network operator to specify. For instance, 'Service Interaction Indicators' in the Connect operation are for the SCF to convey instructions about feature interaction control between IN services and the network-based services. However, this is heavily dependent on switch design and on how the operator wants the services to interact (or not). The CS-1 feature interaction manager (FIM – Section 2.2.10) addresses the interaction of IN service features, but the bigger issues are to do with the interaction of IN-provided services and switch-based 'embedded' network services. Extensive predictive testing of possible interaction scenarios is needed, and if necessary services should be broken down into their component parts and tested individually with network services. As discussed later in Section 4.6.2, part of the CS-2 specification (Q.1224) lays out the principles for IN vs. switch based interactions using the UFM formal methods, but switch-based services are generally so proprietary that in most cases there is still no real alternative to comprehensive testing of individual scenarios.

To allow for different legacy implementations and different operators' particular service requirements, most of the IN operations specified include an 'extension' field. These extension fields provide carriers in the message for items that are operator-specific. For example, the extension fields in the PlayAnnouncement and P&C operations could be used to invoke custom dialogues at an SRF, enabling advanced speech recognition devices and software to be used.

As another example of the use of extension fields, a particular IN implementation may specify that the EventReportBCSM operation should carry a time-stamp for the event in question. This could be achieved using an extension field parameter. Extension fields have proved invaluable in supplementing standard features with operators' specific needs, but they need to be looked at carefully when ensuring interworking compatibility between implementations.

Lower level signalling interface agreements must also be negotiated in detail, because some of the standard is not mandatory. Therefore different vendors may choose subtly different formats, yet still remain within the confines of the standard recommendation. An example of this is the TCAP 'contents length' field (defined in [14]), which specifies three possible forms for this field – 'short', 'long' and 'indeterminate'. The choice of such options obviously needs to be agreed for successful interworking between two IN network elements.

### 2.6.3 SSP data management

The sets of T_DPs, E_DPs and trigger criteria specified in the standard are likely to translate into a non-trivial data provisioning mechanism for the switch, particularly if the new features become successful and many IN services, needing differing triggering conditions, co-exist. Some of the data may be relatively transitory, with some services provided for short trial periods only, but other data will be more durable.

There is potential for different services to interact adversely with each other if the SSP data are incorrectly provisioned. Access safeguards are necessary; it is possible to achieve catastrophic results by misplacing IN trigger data in the SSP. For instance, if someone (inadvertently) inserted an office-based T_DP 1 trigger at a busy time of day, the spectacular result would be that every call on the exchange would trigger before the dial tone. This would result in near-immediate focused overload at the SCP and the exchange itself.

Tight control of the SSP triggering data-builds is therefore necessary, and the importance of easily comprehensible and up-to-date user-guide documents cannot be under-estimated!

*Chapter 3*
# Signalling intelligence

*The wireless music box has no imaginable commercial value. Who would pay for a message sent to nobody in particular?*

> David Sarnoff's associates, in response to his urgings for
> investment in the radio in the 1920s.

## 3.1 Introduction

Whilst this book is primarily about IN we have to remember that IN was introduced into the context of existing voice networks, which already exhibit a fair degree of 'intelligence', both in terms of 'embedded' switch-based services and the common-channel signalling system and the associated call control procedures. Furthermore, without some of these intelligence features already installed in voice networks, the IN enhancements would actually not have been possible.

In order to underpin the discussion on the incremental intelligence features provided by IN we now review the common-channel signalling system and the traditional call processing procedures.

## 3.2 Common channel signalling – the beginning

The introduction of common channel signalling, i.e. the CCITT Signalling System No.7 (SS7), was revolutionary. Until then signalling systems were mainly in-band and line-based, using 'make-and-breaks' on the transmission path, or various combinations of tones on the speech path [15].

Implementations of SS7 started to appear in the early 1980s, along with SPC (stored program control) digital telephone exchanges [16, 17]. The new common channel signalling enabled the switch processors to communicate directly, sending call set-up information between each other using labelled messages. The advent of SS7 allowed the signalling to be removed completely from the speech path, to the enormous benefit of both. Speech circuits could be provided with complete freedom from danger of interference from signalling noise, and SS7 combined the signalling for many speech circuits into a dedicated 64 kbit/s-channel [18].

## 3.3 The layered signalling model

Figure 3.1 gives an overview of the main constituents of the SS7 family of protocols. The basis is a layered model, which has a loose association with the OSI model of the early 1980s.

The diagram shows the relationship between the SS7 model and the 7-layer OSI model (which actually post-dates the SS7 model definition). It can be seen that the match between the SS7 levels and the OSI layers is not exact.

The OSI model was defined by the International Organisation for Standardisation (ISO) to promote open networking environments with the aim of allowing equipment from multiple computer equipment vendors to interwork in a standard manner [19]. Each of the OSI layers is designed to communicate independently with its peer at the distant computer. The lowest layer provides the actual physical bit-stream communications between the two computers, and the next layer up, the

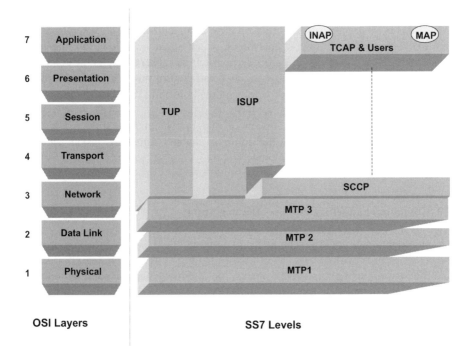

*Figure 3.1   The SS7 layered signalling model*

'data link layer', provides the link control functions such as error control and flow control. Addressing and routing of messages is carried out at the 'network' layer, presenting the 'transport' layer (Layer 4) with an end-to-end network service. Layer 5 is responsible for organising and managing 'sessions' between communicating application processes. In some cases a transaction relationship is needed; in others an extended dialogue is more appropriate – this is the concern of Layer 5. The 'presentation' layer ensures that the syntax of the data exchanged is understandable to each end and the 'application' layer represents the end user in specifying the nature of the required communications. When data is transmitted, each stage, from Layer 7 down to Layer 2, encapsulates the data received from the layer above along with its own header, and passes it to the layer below. Layer 1 takes the data from Layer 2 and sends it bit by bit to the transmission line.

As Figure 3.1 suggests, SS7 signalling evolved from an original 4-layer model, where the underlying message transfer part served just the ISUP, TUP (telephony user part) and TUP-derivatives [21]. We return to these call control protocols in Section 3.5.1.

The signalling model was extended for the protocol requirements of IN and MAP (mobile application part), which both needed reliable and real-time computer–computer transaction-handling communications independent of telephony circuits. The additions to the SS7 model that allow this are SCCP and TCAP, which are discussed later in this chapter.

The foundation component of the SS7 hierarchy is the message transfer part (MTP), which provides the actual messaging service, and is usually regarded as the cornerstone of the SS7 specification set. As indicated in the diagram, MTP has Level 1, 2 and 3 functions (explained in the next section), and can have several users. Some users, such as TUP, employ MTP to transfer 'circuit-related' signalling, where the messages are closely tied to a particular speech (or other bearer transport) circuit. Other users, such as IN, need the SS7 network to conduct computing transactions that are independent of any particular bearer channel. In this latter case, the SCCP is used as an intermediary. We shall return to these higher level components later.

## 3.4 The message transfer part

The purpose of the MTP is to send messages, with guaranteed accuracy and reliability, between user applications across the signalling network. These messages can be delivered directly between two signalling points (SPs), or indirectly by making use of intermediate signalling nodes. The intermediate signalling points are generally known as signalling transfer points (STPs). The SS7 network is often represented as a packet-switched signalling network of SPs and STPs overlaid on the transport network, which physically interconnects the exchanges. This is illustrated in Figure 3.2.

MTP is a reliable packet switching system, where the packets are despatched over synchronous signalling links. In the UK these signalling links are usually

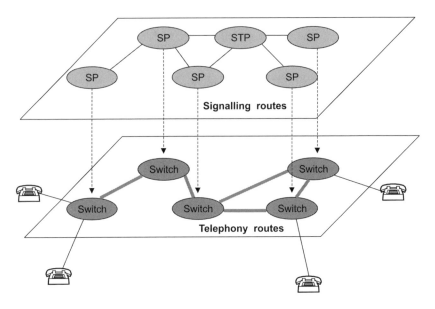

*Figure 3.2   The separated signalling network*

multiplexed into a 32-channel 2 Mbits/s PCM transmission system, typically occupying time-slot 16. For short distances, these 2 Mbits/s systems can be carried over coax cables pairs (one cable each for the 'go' and 'return' directions), but they are multiplexed into higher order systems for transmission over longer distances.

Where a signalling point in the network acts as an STP, its prime concern is to transfer packets from incoming links to the correct outgoing links, in the same way that an IP router switches packets from one route to another. However, if the signalling point also has user applications, such as the call processing software in a telephone exchange, the MTP's main purpose is to provide the signalling for the user processes. A network node can of course perform both these functions, and Figure 3.3 shows the various functions within MTP which make this possible.

The overall purpose of MTP is to transfer messages between users (top left of Figure 3.3) in a controlled manner using the interconnecting signalling links (bottom right). The message routing function ensures that messages from users reach the correct outgoing signalling link. When an MTP user application requires a message to be delivered to one of its peers at a distant node, it will send the information, fronted by a routing label containing originating and destination point codes (OPC and DPC), to the message routing function for packaging and onward transmission [20].

On the basis of the DPC, the message routing function will direct the outgoing message towards the appropriate signalling route and thence to a particular link in that route. At this point the MTP Level 2 procedures will add the forward and backward sequence numbers and checksum information, which is needed to ensure

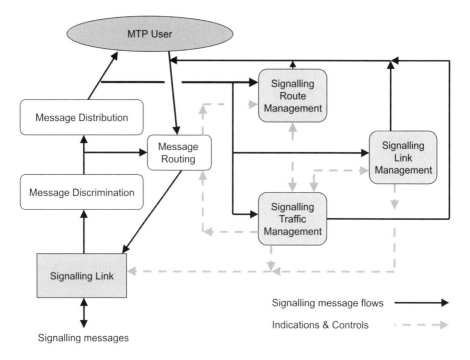

*Figure 3.3   MTP functions*

that the messages are securely delivered in the correct order without loss or duplication.

At MTP Level 1 the signalling links operate synchronously, at 64 kbits/s,[10] and when there is no user traffic to transfer, FISU (fill-in signalling units) are sent. Against this synchronous background, user-signalling traffic arrives spontaneously, and is included in the form of MSUs (message signalling units) in the signalling stream. Link status signalling units (LSSUs) are used to exchange status information between the ends of the link.

For incoming messages, the Level 2 MTP functions will check the sequence numbering and the checksum and will pass the content of the message up to the message discrimination function (shown in Figure 3.3). The message discrimination function checks the DPC of the incoming message. If the DPC corresponds to this node the message is passed up to the message distribution function, which directs it to the appropriate MTP user. The MTP user is determined from examination of a service information octet (SIO). If the DPC does not correspond to this node, and the message is a valid one, then this signalling point is acting as an STP and the message discrimination function sends the message to the message

---

[10] . . . . . . or 56 kbits/s in USA standards.

routing function. The message is then routed toward its required destination using the outgoing routing tables, and is treated in the same way as a normal outgoing message from this SP.

Figure 3.3 also shows the MTP management functions, which, as well as providing the background functions essential for the correct working of the signalling messaging handling, also contribute the major part of the ingenuity of MTP. The signalling management functions ensure that MTP provides a reliable service to its users, even in the face of network failures and disruptions to signalling routes and links. Every signalling point therefore needs to have current knowledge about the reachability of local and distant parts of the signalling network. To provide for this, the signalling management software processes at different signalling nodes send each other information messages to allow their routing data tables to be kept up to date as changes occur.

A signalling point will be aware through the normal Level 2 procedures (see below) when a directly connected signalling point becomes unavailable. If, however, it has a signalling relationship with a more remote node that becomes unreachable, it will receive information about the far end's unavailability through MTP 'transfer prohibited' messages. The node can then take appropriate action, such as re-routing the signalling messages, or refusing to accept more traffic until it receives a 'transfer allowed' message telling it that the problem has been cleared.

The routing data tables provide information on the availability of signalling routes and signalling links. Signalling links are provided in 'link-sets', usually comprising two (for security) or more links, depending on signalling traffic volume requirements. Link-sets are provided co-operatively between signalling points, and on each link they continually exchange signal units, sending FISUs in the absence of MSU payload traffic. The signalling points at each end of the link-set are therefore always aware of the availability of the links. If a link goes out of service there are link management procedures for redirecting traffic onto a good link, and messages which had just been sent on the failed link can be recovered from a buffer and retransmitted on the good link.

Signalling routes are assigned to link-sets, and so if a link-set becomes unavailable then any signalling route assigned to that link-set also becomes unavailable. To provide signalling security, it is common practice to allocate alternative signalling routes for a user traffic route, using intermediate signal transfer points. Signalling routes with a common destination grouped together in this manner comprise a 'route-set', and it is the route-set as a whole that is made available to provide a secure signalling facility to the Level 4 user.

*3.4.1 Signalling management intelligence*

The ability of the signalling network to automatically 're-route' to a second or third signalling route (a collection of signalling links) at times of congestion or subsequent node unavailability gives some insight into the built-in distributed intelligence possessed by the SS7 network. Another, and perhaps more pointed, indication of this built-in intelligence is the ability of the MTP network to distribute information about node unavailability. Each SS7 signalling terminal at an exchange runs a 'watch-dog timer', and if it detects a loss of communication with the exchange central processor, it will autonomously signal 'processor outage' to the node at the far end of that link. The far end will, in turn, inform any surrounding nodes, which might use that node as a signal transfer point, that it cannot transfer messages to the unavailable node.

The existence of this distributed intelligence, automatic signalling re-routing coupled with the call processing software's (normal) ability to automatically re-route calls (independently of any signalling re-routing) sometimes is a source of unease. Implementations need to be sufficiently robust to control the generation of large quantities of automatic management messages that might otherwise swamp the network at difficult times.

## 3.5 Telephony user part intelligence

We now move up from the MTP layers in Figure 3.1. This section illustrates the ISUP and TUP telephony handling functions that use the services of the lower MTP layers.

### 3.5.1 Basic call set-up

The original purpose of SS7 signalling was to enable telephone calls to be set up and cleared down across digital telephone networks. The first major recommendations for SS7 were published by the ITU-T in 1980. These recommendations included the telephone user part (TUP) [21], which provided the framework for the messaging needed to set up, connect and clear down calls.

Whilst TUP defined the common standard, in practice individual network operators defined additions to the standard to suit their own particular network flavours. This was because different administrations had inevitably developed their own way of handling the more complex call types, such as operator calls.

These local variations were called NUPs (National User Parts). However, with the introduction of a multiplicity of network operators in the 1990s, it became important to re-establish the standard for interconnecting public networks. In the UK, an initial pubic telephony protocol for telephony interconnections, known as IUP (Interconnect User Part), was derived from BT's NUP. The latest standard for interconnecting networks in the UK is the UK variant [35] of the ISUP protocol.

As an illustration of how the telephony protocols work, Figure 3.4 shows the SS7 level 4 (call control) messages that are typically exchanged between two local exchange calls using the ISUP protocol in a simple telephone.

**IAM (initial address message)**. This is the first (and the most informative) message of a call set-up. It carries several parameter fields that describe the calling terminal and the call handling required. In particular it carries the dialled address digits, and an implicit request to the next exchange to establish a bearer connection to the required destination. If all the address digits are available at the start of the call set-up (for instance if the complete address has been received in an ISDN set-up request message) then the originating exchange can forward the complete digit string in the IAM. This is known as 'en bloc' signalling, and this is the scheme that is illustrated in Figure 3.4. After the destination exchange is satisfied that it has collected the correct number of digits needed to establish a bearer connection it returns an **address complete message (ACM)**. This indicates to the originating exchange that the destination phone is ringing, and that the switch path can be switched through in the reverse direction to enable the caller to hear the ringing tone over from the far end.

If there is a time lag in sending the digits then an IAM containing just the initial digits can be sent, and subsequent digits are sent in subsequent address messages (SAMs). This is known as 'overlap' signalling and can also be used when the originating node does not know the number length of the complete address digit train – the far end will have this information and can therefore send the address complete message when sufficient digits have been received.

When the called party picks up the phone, an **'answer' message (ANS)** is returned and the originating DLE can switch through the speech path in both directions and start the call charging mechanism. Finally, at the end of the call, one end sends a 'release' message, call charging is stopped and the other end returns a 'release complete message'.

The sending of address digit information constitutes the bearer establishment control phase, which is followed by the call establishment phase when the

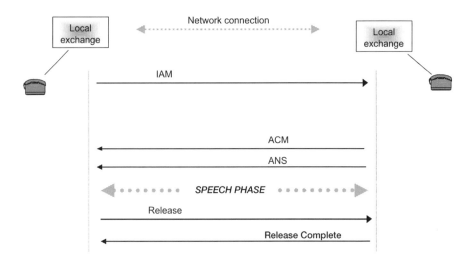

*Figure 3.4   ISUP message exchange for a simple call*

destination needs to determine whether it can support the call protocol requested in the IAM. The messages for a particular call set-up are correlated by a 'circuit identity code' (CIC), which appears in every message for a call and relates to the time-slot chosen in the 30 channel bearer. The IAM includes calling line identity (CLI), allowing the destination DLE to provide services such as 'calling line display' and 'selective call barring'.

### 3.5.2 The 'intelligence' aspects of basic calls

Several 'intelligent' services can be provided in conjunction with a basic call set-up using SS7 signalling without adding any new features. A particular example of this is the set of CLI-based services (see Figure 3.5).

A common feature of SSP telephony signalling systems (ISUP and TUP) is to convey CLI information from the originating exchange to the destination exchange. This is commonly done for administrative reasons, such as accounting, detection of malicious call perpetrators, presentation to a help-desk, etc., but also several supplementary services build on this ability of SS7 to pass CLI information across the network.

#### 3.5.2.1 ISDN CLI services

The supplementary ISDN CLI-based services can apply to 'connected' (i.e. terminating) lines as well as calling lines. A call originator may wish to display the number of the line to which they are actually connected, as this may be different from the number dialled.

Common CLI-based ISDN supplementary services are:

- **CLIP** (calling line identity presentation). This service enables customers to display callers' line identities on incoming calls.
- **CLIR** (calling line identity restriction). This allows the originating end to mask the display of calling line identity at the terminating end, and so it over-rides a terminating user's CLIP service.

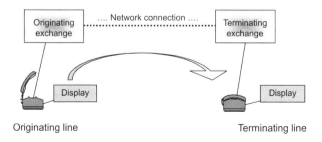

*Figure 3.5 CLI-based services*

- **COLP** (connected line presentation). For outgoing calls this enables users to see a display of the terminating line identity when a call is connected.
- **COLR** (connected line restriction) allows the terminating end to prohibit the display of his line identity at the originating user's display, so over-riding the COLP service if the caller has it enabled.

In Europe the CLI-based services are subject to ETSI standards and guidelines. Interactions between these and other ISDN supplementary services have recently been standardised in Reference 22.

In the case of ISDN, the CLI can be provided either by the network, using data held in the customer data tables at the originating local exchange, or provided by the user in the DSS1 signalling message. The interface from a standard POTS terminal only conveys dialled address digits, so a caller cannot send a 'user-provided' number in the way that an ISDN user can.

### 3.5.2.2  POTS CLI services

For outgoing POTS calls it is common to be able to prevent CLI from being sent forward by dialling a prefix code such as '141'. This is equivalent to the ISDN CLIR service. At the terminating end, it is common for customers to be able to subscribe to a service that enables the CLI to be logged at a display device before it is answered. This is equivalent to the ISDN CLIP service. If the originator has withheld the CLI, 'number withheld' will usually be shown on the display at the terminating end.

However, there are cases where, for technical or regulatory reasons, the CLI is unavailable for forwarding to the destination terminal. The signalling procedures distinguish between this and the case where the caller actually opts to withhold the CLI. Further, a 'presentation' number is sometimes preferred to the 'network' number, which is the public number assigned to a termination. A presentation number might for instance be a Freephone number, which might be a more useful number for customers to call back on than the network number of the originating terminal.

There are other variants on the CLI services. A common feature offered is for the incoming exchange to automatically reject calls where the CLI has been withheld. Call recipients can also interrogate the exchange, by dialling a code such as '1471', about the origin and time of missed call attempts, and then automatically return the call by dialling another code. For calls to busy numbers, a 'ring-back' feature is usually available, where the network can be instructed to automatically connect the call when the destination number becomes available. A European standard version of this service is known as CCBS (call completion to a busy subscriber).

Whilst these features illustrate the intelligence that resides in the SS7 signalling network, they are all in fact made possible by the network signalling procedures, and do not, in their basic forms, necessarily need any IN involvement, although the IN could further embellish the services.

### 3.5.3 Further 'switch-based' intelligence

This transfer of CLI information was one of the main service features of early SS7 TUP implementations. However, because early digital exchanges usually had a set of extra services, and there were slight messaging differences for operator calls, there were several additions and variations to the simple call sequence of Figure 3.4. More recently, services such as 'ring-back when free' were introduced and further signalling modifications were made. Because detailed international standards were not available for these extra services, and because there were different implementations in different administrations, these signalling modifications were often implemented as upgrades to national variants of TUP. Typically, non-standardised values of the service handling protocol (SHP) parameter have been used to instigate more complex messaging dialogues for such services. The parallel development of ISDN and non-ISDN signalling specifications led to differences in terminology and procedures for the supplementary services.

## 3.6 SS7 signalling for IN

So far this chapter has only looked at 'circuit-related' SS7 signalling, where the identity number of the speech circuit is used as the reference key for the signalling interchange. Right from the early days of SS7 design there were plans to also use SS7 for computer-to-computer transactions (with no associated speech circuit) [3], and so provision was made for non-circuit-related signalling for applications such as INAP and MAP.

### 3.6.1 Signalling connection control part (SCCP)

The first CCITT SCCP recommendations were published in 1984 in the so-called 'Red Book'. These were updated four years later, along with significant TUP changes [21], when the C7 'Blue Book' was published. The 'Blue Book' series of signalling standards also included details of the network management aspects of SCCP. The SCCP was designed primarily to support transaction-based messaging using connectionless signalling on the established common-control signalling network, for which it is now used extensively, with many implementations still based on the Blue Book recommendations. However, SCCP standards were completed well in advance of widespread adoption of connectionless signalling and so from an implementer's view much of the detail was still quite theoretical. Anticipated applications for SCCP at the time included [23]:

☐ interrogation of centralised databases by telephone exchange processors
☐ updating of location registers for mobile communications
☐ remote activation of supplementary services
☐ transfer of data between network management nodes.

Of these applications, SCCP is now in extensive use for the first two, and the third has not been standardised and is in limited use in proprietary systems. The fourth was dropped because of the risk of swamping the real-time signalling channels with large quantities of background data, adversely affecting the performance of the network.

Current ITU-T SS7 SCCP recommendations are given in Reference 24.

### 3.6.2 Global title addressing

At the lower level, MTP uses point codes for message addressing, but a major advantage of SCCP is that it can use global title (GT) addressing, which allows greater flexibility. A GT represents a destination user or software process in terms of an agreed format, which could, for example, be E.164 address digits. Before the message reaches its destination it will encounter a GTT (GT translation) function, which will map the GT address to a routable address, which could be an SS7 point code and an SCCP subsystem number (SSN). An SSN is a sub-address representing the SCCP user-function.

Figure 3.6 shows a network scenario using global title translation. A Freephone call triggers from an SSP and there is a choice of SCPs that can handle the request. Because there is an intervening SCCP translator node that can direct the request to

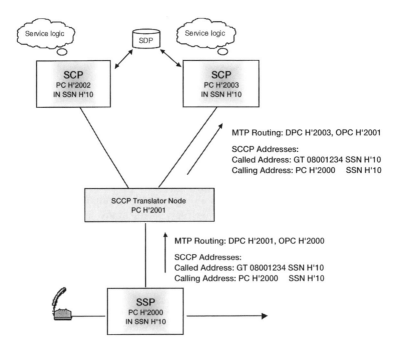

*Figure 3.6    SCCP routing – from SSP to SCP*

whichever SCP is currently the first choice, the SSP does not have to be aware of the network identity of the target SCP. The SSF in the SSP addresses the request to a GT that identifies the service. In this case we choose to identify the service by the most significant eight digits that have been dialled by the caller (historically it is common in many implementations for the dialled digits to be used as the global title identity in this manner). At the MTP level the SSP does not need to have a pre-established signalling relationship with the destination SCP – it just needs to have a signalling route to the translator node, so it routes the message to the translator node's point code.

On receipt of the message, the translator node looks up the called party global title address in its translation tables. In this case the resulting translation is the point code for the first choice SCP and the sub-system number representing IN applications at the SCP. The sub-system number (SSN) represents the user of SCCP. SCCP management messages will be directed to and from a separate SSN, and there are different numbers for other sets of applications, such as MAP services, which may be hosted on the same control point.

When the request message arrives at the SCP the GT enables the request to be routed to the appropriate service logic and a Freephone translation will then be returned to the sending SSP. Figure 3.6 shows the SCPs sharing an external SDP, but this may not always be the optimum architectural arrangement. It will depend on many factors to do with a particular service implementation, such as reliability, speed of response, database size, messaging loading and the capabilities of the database management system.

Figure 3.7 shows the return message from the SCP to the SSP. Because the SCP is now aware of the point code of the originating SSP, it can use this in the SCCP address for the return message. If there is a signalling route configured between the SCP and SSP the SCP can direct the message to the SSP itself (almost certainly via an STP, which may be the translator node itself). However, Figure 3.7 assumes there is no pre-set signalling relationship with the SSP, so at the MTP level the SCP must route the message towards the translator node's point code even though it puts the target SSP's point code in the SCCP called party address field (with the SSN). This time the translator node discovers the SSP point code in the translation look-up, and so it is able to instruct the outgoing MTP to route the message to the SSP's point code.

This example explores one particular routing scenario, in order to explain the global title translation function, but many variations are possible, depending on individual network needs.

### 3.6.3 Transaction capabilities application part (TCAP)

SCCP provides the flexible data transfer mechanism needed for IN, and a powerful routing and addressing capability. To enable transaction-based protocols users such as INAP and MAP to use the services of SCCP, SS7 provides transaction capabilities [26] that interface directly with SCCP. These capabilities are presented to the SSP as the TCAP application. As shown earlier in Figure 3.1, TCAP and its

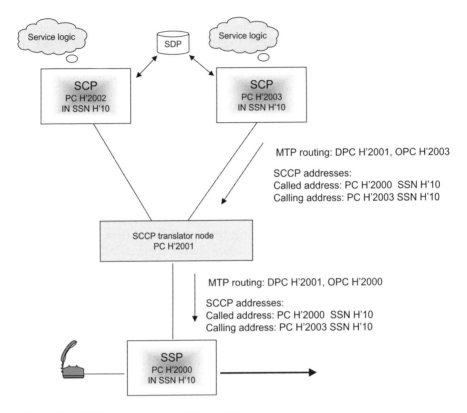

*Figure 3.7   SCCP routing – from SCP to SSP*

users (such as INAP and MAP) reside at the top level of the SS7 stack [25]. This is equivalent to Layer 7 (the application layer) on the OSI stack working directly to Layer 4 (the network layer) without requiring the services of the intermediate presentation, session and transport layers. As Figure 3.1 shows, ISUP and TUP derived protocols implicitly contain these intervening layers.

TCAP's role in IN is effectively to organise and manage the transactions used by the INAP operations and parameters that are exchanged between SSFs, SCFs and SRFs. TCAP performs a similar function for MAP, which enables a mobile switching centre (MSC)'s VLR (visitor location register) to request service profile details from the mobile customer's HLR (home location register).

TCAP messages are carried in the SCCP part of an MSU (an MTP message signalling unit) over an SS7 link. A TCAP message is made up of a transaction portion and a component portion [26]. Information elements in TCAP use 'name, length, value' encoding, which gives TCAP its flexible applicability, although the messaging overhead can be appreciable.

The following sub-sections briefly describe the SS7 TCAP transaction and component portions. However, the signalling mechanism is best described by an example message sequence, and so we illustrate the action of the TCAP elements in

Appendix 3, where we show the sequencing of the messages for a CAMEL mobile network call. CAMEL (customised application of mobile networks enhanced logic) uses the SS7 CAMEL Application Protocol (CAP), which in turn is a user of TCAP, SCCP and MTP signalling in just the same way as INAP. Therefore the behaviour of SCCP and TCAP in transporting the CAP messaging is identical to that for the transport of INAP messages. In fact, as we shall show in later chapters, there is close equivalence between CAP and INAP, because – as we shall see – the former was derived from the latter.

### 3.6.3.1 Transaction portion

The transaction (or dialogue) portion of TCAP co-ordinates and correlates messages using transaction identities. This allows the existence of multiple transactions between end-points. A message can have an originating transaction identity (OTID) and a destination terminating identity (DTID), but not necessarily both, and the messages are grouped into one of four types:

◇ **Begin**          This initiates a transaction with a remote node.
◇ **Continue**       This continues an existing transaction.
◇ **End**            This is a normal termination of a transaction.
◇ **Abort**          This is an abnormal termination for a transaction.

The transaction Id. is locally assigned by the sending node. The 'continue' message type is the only one that has both OTID and DTID. The 'begin' will only have OTID and the 'end' and 'abort' message types only have the DTID.

### 3.6.3.2 Component portion

The component portion of TCAP contains one or more components and is responsible for correlating responses to requests and handling error detection. The components are based on the application protocol data unit (APDU) concept of the ITU-T ROSE (remote operations service element) defined in ITU-T X.229.

Components are grouped into one of five types:

| | |
|---|---|
| **Invoke** | This requests the invocation of an action at the remote end. Invoke components have an invoke identity and contain an operational code and (optionally) related parameters, which for ITU-T IN are the INAP operations and parameters described in Chapter 2. |
| **Return result** – **'not last'** | This contains a response to an earlier Invoke. |
| **Return result** – **'last'** | This contains a final response to an earlier Invoke. |
| **Return error** | This indicates that the associated invocation was not successful, and an error code and further error information is provided. |
| **Reject** | TCAP, or the TC user application, sends this when previously sent information is not understood. |

### 3.6.4 Signalling differences between PSTN and GSM networks

Historically the technical differences between fixed and mobile networks have been big enough to justify the evolution of separate network structures. As a result, separate switching network methodologies have established themselves in the fixed network and mobile worlds.

However, in the core of both networks there is commonality in the use of ISUP (or TUP-based) SS7 for interchange of call set-up information between switching centres. INAP is also found in both network types, but in the case of mobile, equipment vendors have generally supplied the network operators with tailored proprietary variants to cater for the networks' particular needs. However, where CAMEL applications are used, the standardised CAP interface is also used, and this has introduced opportunities for multi-vendor interworking of equipment.

Mobile networks generally make richer use of the various SS7 protocols. As well as using ISUP for call set-up signalling, CAP (and proprietary INAP versions), for IN access, SS7 has several other uses in mobile networks.

SS7 MAP is used extensively for the mobility functions between MSCs, HLRs and VLRs. In the 'A interface' between the switching centres and the base station controllers, the BSSMAP manages resources and connections, MAP relays supplementary service information and DTAP carries the radio interface messages (RIL 3). Figure 3.8 compares the variety of SS7 protocols used in mobile calls compared with fixed line calls.

Apart from the obvious differences in access technologies, mobile call processing is more complex because authentication and mobility management

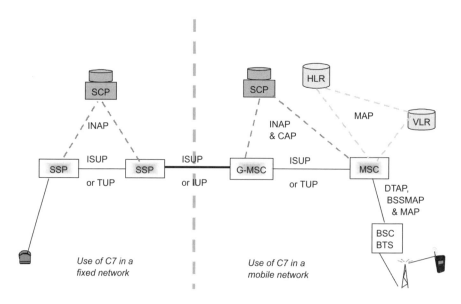

*Figure 3.8   Comparison of PSTN and mobile network signalling architectures*

procedures are needed in addition to the call control function. There are therefore more signalling messages needed for a mobile call than for a fixed network call. This is the reason why mobile phone users have become used to waiting several seconds between dialling a number and hearing a call progress tone (such as ring tone). For a fixed network phone this waiting time is of course typically well under a second.

Of course, technical differences are not the only reasons for the differences between fixed and mobile networks. Business drivers for mobile phones have generally been more urgent in the race to compete with fixed network operators and so less time is available for in-depth analysis of technical options, which used to be normally expected with the incumbent fixed-network operators. In more recent years, speedy evolution has been essential in the race to win customers and so there has been little incentive to seek a common evolution path with fixed networks.

As for intelligent network developments, Figure 3.8 shows two separate IN SCPs. This is because, during the formative years of mobile networks, there was little effort made to converge the mobile IN service control functions with those in neighbouring fixed networks. Mobile network evolution paths were driven primarily by the procurement of vendor implementations that could deliver the required service features. Consistency across fixed-network and mobile technologies was rarely a significant factor in procurement decisions on either side of the fence, because operators tended to specialise in either fixed or mobile networks, and so were not overly concerned with the other side.

Different switch designs have necessitated different triggering arrangements, and the use of different IN standard variants has ensured that services are generally implemented differently in the SCPs.

# Chapter 4

# International standards for intelligent networking

*While considering which of two things you should teach your child first, another child has learned them both.*

Dr Samuel Johnson (1709–1789)

## 4.1 Introduction

IN was once heralded as a network re-arrangement that would provide any telecommunications service need that any user could conceivably want. In practice, over the years, commercial priorities, financial constraints and inter-working limitations have inevitably limited the vision's potential. Whilst in the early days the IN concept had high aspirations, reality had to intervene sooner or later to remind us what might be feasible and what might not be. For example, call routing decisions based on the time of day are perfectly feasible, whereas a routing decision based on the colour of a user's shirt may be less feasible!

As already discussed, AT&T and then Bellcore, in the USA, made the running on early exploratory work on IN specifications and the European operators watched the results carefully. In the early 1990s there was a lot of activity in the USA on IN specification production, although these standards being formulated by Bellcore were obviously designed to meet US needs and US switching system requirements, some of which were less relevant outside the US.

In the early 1990s, the CCITT (now the ITU-T) set up a study group hierarchy for intelligent network standards, and the first result was the CCITT IN CS-1 (Capability Set 1) in 1993. CS-1 was planned to be the first of several steps towards the introduction of intelligent network concepts into telephony networks. It was seen as a relatively modest set of early IN capabilities for voice calls. It deals with single-ended, single point of control, IN service features.

'Single-ended, single point of control' means that there is just one controlling SCP acting on a call for any one service invocation. What it excludes is the possibility of having co-operative service invocations, controlled by different SCPs and acting on different SSPs, affecting the same call. The important point is that all the invocations must be independent – there is no dialogue between separate SCPs about the same service invocation. This means that the SSF is not able to interact with more than one SCF at any given time in order to complete a sequence of queries and responses (forming an IN transaction) on behalf of a calling or called party. As we cannot have more than one SSF–SCF relationship taking part in an IN transaction an SCP is unable to orchestrate a call across a network by simultaneous SCF–SSF transactions. Figure 4.1 shows two separate SCP invocations on one call, which is allowable. These could, for instance, be a user authorisation service, where a check on a caller's identity is made with the first SCP followed by a request to a second SCP later in the call for a 'least cost routing' service. The two SCPs are working in isolation and so the 'single-ended, single point of control' rule is not broken.

The use of the term 'SCP' is representational here. Strictly the 'control point' is a single computer housing the requisite service logic. In fact, as we shall see later, the service control functions may be distributed geographically. Also, in the stricter analysis to follow, the term 'service control function' (SCF) is used, indicating that the control functionality can be hosted at a variety of locations – not necessarily in SCPs.

Another important limitation for CS-1 is that it is really for two-party calls only, and there are no IN functions provided that will allow manipulation of conference calls involving three or more users.

In the early ITU-T standards deliberations there was a lot of discussion about the form of the IN call model that should be included in CS-1. In particular, some delegations did not want CS-1 to include some of the more advanced ideas of 'leg-control' (which had already been put forward in the Bellcore IN/2 proposal in the US but had not been received favourably by switch vendors). Leg control enables IN service logic to manipulate the different parties involved in a call, enabling provision for IN controlled conference calls. Whilst the ITU-T CS-1 contents were

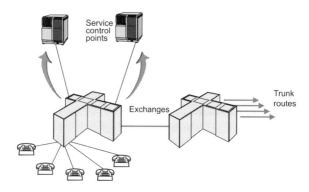

*Figure 4.1   Separate service invocations on the same call*

being decided, Bellcore was standardising the AIN set of recommendations in the US in parallel with the ITU activities, although there was co-operation between the two activities.

## 4.2 Early US standards for IN

In the US, the catalyst for early rapid IN development was the '1800' Freephone service. IN really originated in the middle of the 1980s, when the seven US regional Bell operating companies (RBOCs) expressed interest in centralising the service control for services such as Freephone and Calling Card. Because the US had retained letters on telephone dials and key-pads, entrepreneurial ingenuity was ensuring the success of the 1800 service. Advertisements such as 'dial 1800 FLOWERS' (for flower deliveries) were becoming common. The calling card verification service, also known as 'alternate billing', enabled users to make calls from public payphones and have the call charges accrue to an account independent of the normal exchange charging mechanisms.

More recently there have been a series of obligations imposed by the US Federal Communications Commission (FCC) concerning Local Number Portability (LNP). The 1996 Telecommunications Act demanded that the US local exchange carriers (LECs) must allow customers to take their phone numbers with them when they moved from one carrier to another whilst maintaining existing standards of service quality and reliability. This mandate was subsequently extended to include wireless carriers. Much of the recent investment in AIN implementations has therefore been to meet these requirements.

### 4.2.1 INWATS, IN/1 and IN/2

The early AT&T toll carrier exchanges in the 1970s were designed to offer an important feature known as Inward Wide Area Telecommunications Service (INWATS). This was the original Freephone service. The switches were designed to route the '1800' service request messages, using the new common channel signalling system, to a central database, which would translate the digits dialled by the caller into a standard NANP (North American numbering plan) routing address and return it to the controlling switch. This was followed in the early 1980s by the addition of the Calling Card verification service using similar network principles. Evolution of these principles at Bell Labs led to the DSDC (direct services dialing capabilities) architecture. The DSDC access switches were known as ACPs (action control points) and the central database was known as the NCP (network control point). Bell Labs also added the NSCX (network services complex) for DTMF and voice interaction with the caller, and this became the forerunner of the IN Intelligent Peripheral.

A system based on the DSDC, from AT&T Network Systems, was installed by BT in the UK in 1988, where it became known as the derived services network

(DSN). The DSN provided the basis for BT's Freephone service for several years. Similar systems were installed elsewhere in Europe in the early 1990s.

In 1984 the giant Bell company was broken up, and AT&T, the toll carrier company, became separated from the Regional Bell operating companies (RBOCs) who owned the local switching networks and customer access switches. As a result of this, the DSDC technology could not be applied to local switches, and the application of DSDC-type functions became an important sphere of interest of Bellcore (Bell communications research – since renamed Telcordia Technologies). Bellcore began to use the term 'intelligent network' for providing DSDC-like services in the local carrier context.

That period in history was a time of fast growth in the uptake of these additional services, and there was a lot of interest in the market. As well as the RBOCs' interest in networks that could efficiently handle them, there was pressure, through the regulators, from external service providers who wanted to be able to offer their own variants of these enhanced services over the network operators' established telephone networks to their customers.

In response, Bellcore developed the IN/1 and IN/2 architectures. IN/2 was well ahead of its time, and it was not well received by the equipment vendors because it was too wide-ranging. It included advanced leg-control features such as 'split' and 'join', which appeared several years later in ITU-T IN CS-2 specifications, but even today these remain largely unimplemented.

On the other hand, IN/1 (Figure 4.2) was relatively limited, but more realistic in that it could be implemented in practical time-scales to meet the market demand. It was designed for a small set of specific services, whereas IN/2 was generic. IN/1 introduced software upgrades to the network switches to allow them to use the existing SS7 STP network to send a request to an SCP when they detected a user-request for one of the 'special' calls such as Freephone or Calling Card.

As Figure 4.2 illustrates, the characteristic of IN/1 was that it was service specific. In other words it was built around particular services. Dedicated 1800 and alternate billing triggers were implanted in the switches, a separate SCP was needed

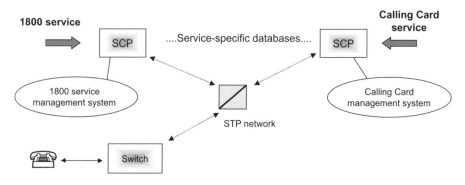

*Figure 4.2   Bellcore IN/1*

for each service, and separate service creation, management and provisioning hierarchies were needed for each.

### 4.2.2 AIN

After the IN/1 and IN/2 initiatives the next set of IN standards from Bellcore was termed 'advanced intelligent network' (AIN). The AIN specification aimed to move away from IN/1's reliance on service-specific components that could not be re-used for other services. The AIN proposal, illustrated in Figure 4.3, was comprehensive and wide-ranging, calling for generic management systems, control points and switch software for triggering. AIN included an IN IP for voice announcements and voice/DTMF recognition resources. The IP could also enter into dialogue with the end-user, on behalf of the service logic at the SCP, in order to collect more information from the user. The IP connected with the SSP using ISDN primary or basic rate interfaces. Also included was an 'adjunct', which provided SCP-like functions but was directly connected to the SSP over a high-speed interface rather than SS7.

However, like the earlier IN/2 proposal, the AIN Release 1 specification was too ambitious for vendors to use to meet the prevailing short-term demand for IN services. Compared with the ITU-T specifications, it was a superset of CS-1, including mid-call trigger detection points and multi-party handling. For this reason phased releases of AIN were established. Bellcore defined AIN 0.1 (broadly equivalent to ITU-T CS-1) as an initial scaled-down set of functions in order to provide an implementation target. In practice the RBOCs set about to define sub-sets to achieve their individual needs, and this resulted in AIN Release 0. AIN Release 0 implementations consisted of proprietary specifications that differed between the RBOCs.

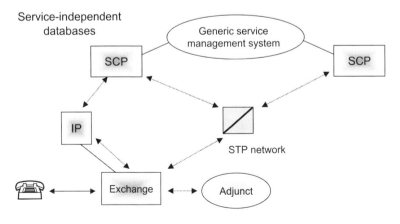

*Figure 4.3   Bellcore AIN*

## 4.3 ITU-T IN CS-1

Whilst the pioneering work was in progress in the US, and regional standards were being formed, the ITU-T began to take an interest in IN. General ideas about the sort of hooks and the sort of services that could be provided were already formulated and in 1989 the ITU-T started to develop the IN vision, which was so successful in the US, into a potential international standard. The concept of providing 'hooks' in traditional telephone exchange call processing software in order to invoke the real-time assistance of external service logic was progressed and developed into the bottom-up, top-down mechanism illustrated in Figure 4.4. Within the scope of what was initially likely to be possible in terms of network upgrades, a set of benchmark services was drafted. A framework process was then assembled in which a protocol could be produced for the first tranche of the ITU-T IN standard.

This 4-layer framework, on the right-hand side of Figure 4.4, became known as the intelligent network conceptual model (INCM), and it provided a backdrop against which to co-ordinate the different IN working groups in the ITU-T whilst components of the overall picture were being assembled. The goal of the entire exercise was to achieve an interworking standard against which network operators could procure IN components and upgrade their networks. The first release of this standard was called IN Capability Set Number 1 (IN CS-1)[27].

The term 'capability set' is actually used as a generic term throughout ITU-T standards, although in this book we use the term 'CS-1' to refer to the first ITU-T IN standard. There is generally not a connection between the capability set versions across the different standards disciplines. For instance, at around the same time as the IN CS-1 recommendation was being assembled a parallel ITU-T study group was forming the initial release of the broadband ISDN (B-ISDN) recommendation. This was also called CS-1. The B-ISDN activity was concerned with converting the

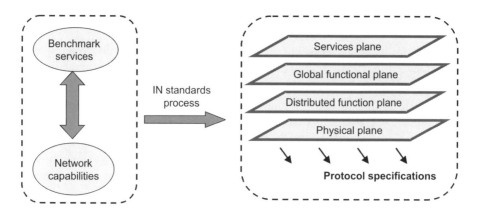

*Figure 4.4   The IN standards-generation process*

existing ISDN point-to-point services to run over ATM, and it was generally based on the familiar narrowband ISDN (N-ISDN) Q.931 standard.

### 4.3.1 IN CS-1 documents

CS-1 was intended as the first of a series of capability sets, and so, in order to indicate an evolution path, it was set against a background of generic recommendations. The ITU-T documentation authorities set aside the Q.12xy block of numbers for IN, where the 'x' digit was allocated to the capability set number, with the value 0 for the generic range of IN documents. The INCM was fundamental to the documentation structure, and Table 4.1 below summarises the documents that were allocated for the generic range.

The CS-1 specific recommendations were numbered as shown in Table 4.2.

We now briefly summarise the contents of these CS-1 documents.

*Q.1210* shows the organisation structure of the CS-1 document series, providing a summary overview for the individual recommendations.

*Q.1211* describes the scope of the CS-1 standard, outlining the network functions and their relationships. It also describes a set of 'benchmark' services that were used for the IN CS-1 recommendation formulation process. These benchmark services are individually described in Section 4.3.2.

*Q.1213* gives details of the 'global functional plane' (GFP), which translates service requirements into the top-level theoretical structure for IN. Part of this process is the definitions of a set of service independent 'building blocks'.

*Q.1214* is a substantial document that shows how the high-level functions of the GFP are distributed into the various IN functional entities (FEs). It describes the operation of the IN CS-1 state model and the resulting information flows.

*Q.1215* shows how the IN FEs can be populated into different physical entities.

*Q.1218* describes the INAP protocol details.

*Q.1219* is a detailed guide to the understanding and the implementation of IN CS-1. It includes some example IN service scenarios.

The set of CS-1 documents was approved for publication in March 1993. Appendix 1 provides a list of all the ITU-T IN documents, with individual publication dates, which were in force at the time of publication of this book.

*Table 4.1   ITU-T IN generic documents*

| | |
|---|---|
| *Q.1200* | General Series Intelligent Network Recommendation Structure |
| *Q.1201* | Principles of Intelligent Network Architecture |
| *Q.1202* | Service Plane Architecture |
| *Q.1203* | Global Functional Plane |
| *Q.1204* | Distributed Functional Plane |
| *Q.1205* | Physical Plane |
| *Q.1208* | General Aspects of the Intelligent Network Application Protocol |

*Table 4.2   The ITU-T IN CS-1 documents*

| | |
|---|---|
| *Q.1210* | Q Series Recommendations Structure for IN CS-1 |
| *Q.1211* | Introduction to IN CS-1 |
| *Q.1213* | Global Functional Plane for CS-1 |
| *Q.1214* | Distributed Functional Plane for CS-1 |
| *Q.1215* | Physical Plane for CS-1 |
| *Q.1218* | Intelligent Network Application Protocol for CS-1 |
| *Q.1219* | User Guide for CS-1 |

In October 1995, an update to CS-1 was issued. This was called CS-1R (for 'refined'). It incorporated solutions to issues raised by industry during the process of defining the ETSI Core INAP standard. CS-1R is discussed further in Section 4.4.

*4.3.2  IN 'benchmark' services*

To establish the formal starting point, a set of 'benchmark' services was agreed and documented in Q.1211. These services were assembled as a list of examples against which to test the IN constructs as they were developed. It must be stressed that these are examples only – IN is not service-specific and it provides the infrastructure for many other services and service variations. Many of the services on the list can (and often are) provided using CPE-based functions, or switch-based software on the PSTN, and they do not necessarily need the extra features of IN. However, as explained earlier, IN provides the opportunity for making significant efficiency improvements by centralising the control, data and management for these services.

Q.1211 contains the precise definitions for the benchmark services. The following comments summarise the essence of the listed services. Services with close similarities have been grouped together in these descriptions.

**Abbreviated dialling**. This service allows commonly used address digit strings to be represented by one or two digits, which can be dialled instead of the complete digit string to reach the required destination address.

**Account card calling** and **automatic alternative billing**. These are variations on the same service, enabling callers to bill call usage to their own accounts or their employers' corporate accounts. The service variants allow the user to either dial the card number and PIN as prefix digits before the required address digits, or to 'swipe' the card through a reading device attached to the terminal equipment. The cost of the call is thereby debited to the user's account rather than to that of the renter of the telephone line. This is particularly useful for making calls from pay phones without having to use coins. The **credit card calling** service extends this capability to include direct billing to a bank credit card.

**Call distribution**. Incoming calls to a customer can be distributed to various answering points depending on the subscriber's own pre-set rules. The **destination**

**call routing** service allows the calls to be directed to different destinations depending on conditions such as time-of-day or call-origination point. The **call re-routing distribution** service additionally allows the call to divert to a different destination if it encounters a problem such as congestion in trying to reach the customer.

**Call forwarding**. Incoming calls can be forwarded to a chosen alternative destination. If the service is provided using IN, a mechanism could be provided for the customer to set up the call forwarding parameters automatically under IN control. **Selective call forwarding** (on busy or no reply) is a refinement where calls are forwarded to an alternative destination if the customer's line is engaged, or there is no reply within a preset time-out period.

**Completion of calls to busy subscribers (CCBS)**. The network automatically arranges for a call that does not succeed (because the called party is engaged) to be retried when the line becomes free.[11]

**Conference calling**. Several users can take part in the same conversation.[12]

**'Follow-me' diversion**. Customers should be able to invoke call diversion from wherever they are so that they can receive their calls on any convenient phone.

**Freephone service**. This is the most well known of all IN services and is subject to much discussion in other parts of the book. It effectively provides a reverse-charge call for anyone who knows the Freephone customer's number.

**Malicious call identification (MCI)**. This is a service that is offered to users who complain of being subjected to abusive incoming calls. If the user invokes the service (by pressing the recall button or a key on the key-pad) then the network operator is alerted. The calling line identity is then recorded and the line held for tracing if necessary. This is another service which, although it was included on the potential list of IN service examples, was available long before IN and so is traditionally provided using switch-based and network signalling functions.

**Mass calling service**. This typically applies to large corporate Freephone customers where the target destination numbers may need many thousands of ports to handle high volumes of incoming calls. An essential feature is the availability of statistics to enable the customer to manage and plan the call reception points.

**Originating call screening**. Before an outgoing call is completed a profile is checked to ensure that the renter of the phone allows the type of call which has been requested. If the call is disallowed, maybe because of its cost, a rejection announcement or tone is played. Conversely, **terminating call screening** can be applied to incoming calls, where calls can be rejected if specified criteria are not met. For instance, calls from a particular CLI could be rejected.

---

[11] *Note*: whilst CCBS was incorporated in the list of benchmark services as a challenge to the CS-1 standards-makers, it actually needs some IN functions which are more advanced than CS-1. In the meantime the service is therefore commonly implemented using non-IN (switch-based) functions, with proprietary network signalling messaging sometimes used to inform the caller when the line becomes free.

[12] *Note*: as with CCBS, it is not possible to provide IN controlled conference calling without more advanced, post-CS-1, IN features. Again, conference calling is a common PSTN service, generally provided with non-IN capabilities until CS-2 features become available.

**Premium rate service**. The premium rate service provides access to sought-after information for which a high charge can be levied. The charge is recovered from the user's phone bill and the revenue is apportioned between the content provider, the service provider and the network operator as appropriate. Distinctive number range codes should be reserved for premium rate services so that a user is aware of the potential high cost before making a premium rate call. The revenue apportionment of premium rate services is an example of the **'split charging'** feature, where IN charging functions can be used to allocate the cost between calling and called parties (as well as to third parties, where appropriate).

**Security screening**. Security screening can be applied at an SCF to check that requests made to a service provider's application have originated from authorised sources.

**Televoting**. IN functions can be used to count the number of calls to particular network addresses that have been previously advertised as options for voting. Callers dial the numbers advertised for their particular voting preferences and calls are registered in the counting service logic. A unit fee is typically raised and the calls are terminated on announcements that assure callers that their votes have been registered. The specific IN network 'activate service filtering' feature, devised for this service application, has been introduced in Section 2.5.1 and is illustrated in a service example later, in Chapter 7.

**Universal access number**. This is sometimes known as a 'personal number' service, where a customer can be allocated a 'personal' number that is separate from the network termination port address (telephone number) of a particular telephone line. This feature allows the **universal personal telecommunications** service to move between different fixed (and mobile) lines and still be reached. The customer can control how incoming calls are handled so that calls can be directed to different network addresses at different times of the day. Calls can automatically be sent to answering machines at certain times, data calls can be sent to data terminals and fax calls to fax terminals. Other criteria, such as the CLI of the caller, can be used to determine where to terminate the call.

**User-defined routing**. A customer is able to specify the network routing to be used for outgoing calls. This is commonly used to express preference in the selection of a particular long-distance carrier network for a call. If the first carrier is not available then the network should attempt the call via a customer-selected second choice.

**Virtual private network** (VPN). This allows a corporate customer to set up what appears to be a private network using public network resources. Customer benefits include a private numbering plan, using short numbers, and reduced call charges to other 'on-net' destinations.

### 4.3.3 The IN conceptual model

The INCM, which was shown in the overview in Figure 4.4, is shown in greater detail in Figure 4.5. Essentially, the analysis starts with the service examples going into the top of the process with the completed protocol specifications, meeting all of the network needs of the example services, coming out of the bottom.

### 4.3.3.1 The services plane

At the top level is the services plane, where services are described in terms of sets of service features. Working down from the 25 'benchmark' services described in Section 4.3.2 above, CS-1 defined [27] a set of 38 separate service features that could support these example services. It must be remembered of course that the purpose of this set of service features was not just to enable the 25 benchmark services to work, but to provide the ingredients for a far wider range of IN services.

The actual list of service features defined for the ITU-T CS-1 is not reproduced here, but as an example, an 'alternate billing' service implementation might make use of (at least) the following three service features:

**Originating user prompter** – this feature is to connect an announcement to the calling line in order to invite the caller to enter extra information, such as a PIN code, using DTMF digits or speech.

**Authorisation code** – this is the ability of the network to collect and process the PIN code when the caller provides it.

**Call logging** – this prepares a call record to be prepared for each call. The record can then be relayed to a billing centre for processing.

### 4.3.3.2 The global functional plane

The plane below the top-level services plane in Figure 4.4 is the global functional plane (GFP). Here, the logic of the service features is expressed in terms of the pre-

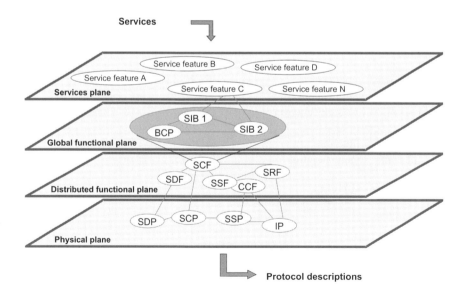

*Figure 4.5   The IN conceptual model*

defined service independent building blocks (SIBs). Just as the GFP is an abstraction of the complete IN-structured network as a single entity, the SIBs are abstract representations of the network's capabilities.

CS-1 defined 13 SIBs [28] for the previously defined service feature set. The SIBs are listed in Table 4.3. CS-1 also defines a 'special' SIB called the basic call process (BCP), whose purpose is to provide an underlying representation of the call control 'thread'. The BCP co-ordinates the invocations of the other SIBs.

Figure 4.5 illustrates how SIBs can be chained in order to emulate service features. SIBs are independent of both the services they are creating and the underlying implementation technology.

Whilst the GFP was part of the overall standards-making model, the anticipation was that the arrangement of standard SIBs in the GFP could be re-used as a standard tool for service designers to create service features from standard network components using a standard programming environment. This has been partially successful, although SIB definitions are high level and there are differences between different implementations of service creation systems. For this reason multi-vendor working involving SCEFs and other IN FEs has not generally been achieved.

We do not pursue the theory of SIB chaining and the GFP any further. Service creation was discussed in some detail in Chapter 2 but for more information the reader is referred to Reference [29].

### 4.3.3.3 The distributed functional plane

The distributed functional plane (DFP), which is the next plane down in the INCM (Figure 4.5), breaks the SIBs down into functional entities (FEs). Six FEs are modelled in the CS-1 DFP; they were all introduced in Chapter 2 and are listed below:

◇ Call control function                (CCF)
◇ Service switching function           (SSF)
◇ Specialised resource function        (SRF)
◇ Service control function             (SCF)
◇ Service data function                (SDF)
◇ Call control agent function          (CCAF).

*Table 4.3   IN CS-1 SIBs*

| | | | |
|---|---|---|---|
| ☐ | Algorithm | ☐ | Screen |
| ☐ | Charge | ☐ | Service data management |
| ☐ | Compare | ☐ | Status notification |
| ☐ | Distribution | ☐ | Translate |
| ☐ | Limit | ☐ | User interaction |
| ☐ | Log call information | ☐ | Verify |
| ☐ | Queue | | |

CS-1 models these six FEs [30], but additionally it describes another three FEs, in preparation for more detailed treatment in later capability set releases.

The CCAF, which represents a network user device that is more sophisticated than an analogue POTs phone, does not trigger directly to an SCF, but conducts all communication through the CCF.

The additional FEs, described by but not modelled by CS-1 are:

◇ Service management function          (SMF)
◇ Service management access function    (SMAF)
◇ Service creation environment function  (SCEF).

These extra functions have already been introduced in Chapter 2. The complete set of the functional entities prescribed by the ITU-T for the generic DFP are described in Reference 40.

Figure 4.6 reproduces the Q.1214 CS-1 DFP, showing the FE relationships applicable for IN CS-1. The two sets of SSF/CCF combinations have identical functionality but are shown because it is possible for one SSF to act as an 'assisting SSF' – where the 'Assist' procedure is used in the same way as it is with an assisting SRF. There are therefore two possible SSF – SCF INAP relationship types.

The FEs (the 'network 'objects') communicate with each other using IN 'information flows' (IFs). An IF is a message that passes between two FEs. Most IFs in the DFP map directly onto operations in the physical plane, and have similar names in both planes.

### 4.3.3.4 The physical plane
The physical plane contains the mapping between the FEs of the previous section – the DFP and physical entities. The mapping is illustrated in Figure 4.7. Seven of the

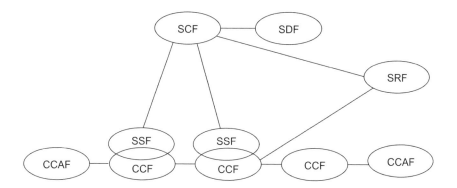

*Figure 4.6   IN distributed functional plane model for CS-1*

Reproduced, with permission, from Figure 3-1/Q.1214 ITU-T publication, 'Distributed Functional Plane for Intelligent Network CS-1' [10/95].

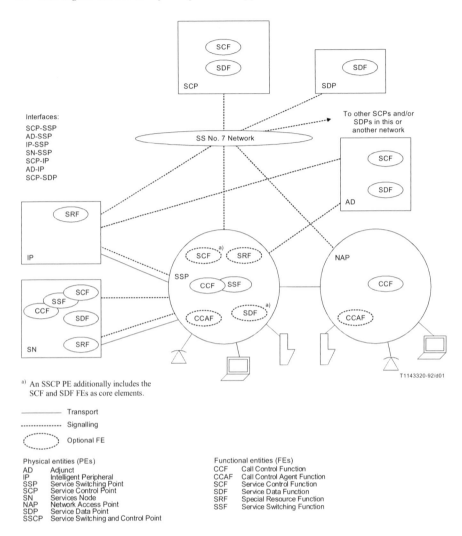

*Figure 4.7    IN physical elements for CS-1*

Reproduced, with permission, from Figure 1/Q.1215 ITU-T publication 'Physical Plane for Intelligent Network CS-1' [10/95].

possible physical entities (PEs) are shown, and some of the FE mappings are optional. It is possible (as indicated in the diagram) to include SDF and SCFs in an SSP, in which case it is normally called an SSCP (service switching control point), which was introduced in Chapter 1. These mappings are representational and other PE–FE mappings are not excluded. It is also possible for FEs to be located in PEs other than those illustrated here [30].

Except for the NAP and AD, the physical entities shown in Figure 4.7 were introduced in Chapter 2. The NAP contains only CCAF and CCF functions and has no means of communicating with an SCP. The AD (adjunct – which was introduced earlier in the chapter in connection with the Bellcore AIN standard) is equivalent to an SCP but is more closely coupled to the SSP, using a higher speed connection than SS7.

## 4.4 ETSI core INAP and CS-1R

After the ITU-T defined CS-1 in 1992 it was found to be too wide-ranging and incomplete in detail to use for detailed implementation. ETSI therefore set about extracting and defining a pragmatic sub-set that could be used as a basis for IN installations in Europe. This work was completed in 1994 and was called the 'core INAP' standard. It was then adopted by the ITU-T and released as an international standard in 1995 under the title CS-1R.

As an indication of the difference between the original ITU-T CS-1 and CS-1R, CS-1 initially defined 55 INAP operations, but only 29 survived in ETSI core INAP. These core INAP operations were described in Chapter 2, and the remaining 26 omitted from core INAP are listed and described in Appendix 2.

## 4.5 IN and mobile systems

An area where IN is having a significant impact today is in mobile telephony. Several standard supplementary services are available in GSM, including features such as conditional call forwarding and call barring, but until the past few years there has been little standards support for some of the more advanced voice services, such as VPN. This has resulted in the development of a variety of non-standard solutions tailored for particular services. Whilst these sorts of solutions are optimised for particular sets of requirements, evolution without agreed standards can obviously be problematic and expensive. Another difficulty which has been emerging with operator-specific solutions for advanced services for mobile users, and one which is becoming increasingly significant, is that familiar services do not always work correctly for international travellers who are temporarily served by a foreign mobile network.

### 4.5.1 The role of IN with GSM

To help with these problems, GSM Phases 2 and 2+ included standard specifications for sets of supplementary services. However, tight standardisation presents operators with a converse dilemma in that if all operators can offer the same services then there is limited scope for competitive differentiation.

For this reason manufacturers started to offer new mobile switches and upgrades that are also able to function as intelligent network SSPs, based on the CS-1 functions. The incentive for introducing IN functions is to give mobile operators the potential for implementing centralised advanced services at reduced costs, and with the flexibility to customise standard services with extra features to distinguish their particular product set.

An option for extending the set of available services, beyond that provided by switch-based signalling procedures, is to provide access from the MSCs to an IN service control point. Hence, some advanced services have been provided in mobile networks by using INAP. However, because the ITU-T IN standards were designed for fixed networks they do not transfer directly to mobile networks. For this reason, proprietary extensions to the INAP protocol, and individual adaptations and modifications of the IN standard network element concepts, were needed, and so inevitably a variety of 'proprietary' INAPs were introduced.

### 4.5.2  CAMEL

In order to make IN solutions available in the mobile world in a more standardised fashion, the GSM CAMEL specification was published. The prime driver for CAMEL was to give mobile network operators the ability to offer their own sets of advanced services to their customers even when roaming into other operators' networks.

CAMEL originated in ETSI and represents the first co-ordinated approach of the IN standards into the world of mobile networks.

CAMEL is the leading IN standard for GSM. Because it originated from ETSI it is derived from the ITU-T IN CS-1 specifications. There is therefore a degree of compatibility between ITU-T fixed network IN systems and GSM CAMEL systems, although they have developed separately to a large extent, so inter-working between the two is not entirely straightforward. Phase 1 of CAMEL (published in 1997) provided a basic set of IN capabilities. The aim was for the interface between the GSM SSP (termed a 'gsmSSP'), which is an upgraded MSC, to follow the ITU-T CS-1 INAP specification as closely as possible, even though the MSC procedures relating to the IN messages are mobile-specific. Figure 4.8 shows the GSM network architecture enhanced with the IN elements.

CAMEL Phase 2 (1998) added the network features needed to support interaction with SRFs (specialised resource functions). These are provided in the IN intelligent peripheral (IP) node, illustrated in Figure 4.8. This allows announcements to be provided, for example, during Pre-pay services, so rather than cutting off a call with no warning when the caller's credit has expired, a suitable announcement can be played to the caller first. Examples of the Pre-pay mobile service are given in Chapter 7.

The IP may be located either in the home network or in a foreign network. For international travellers served by a foreign network, access to an IP located in the same foreign country can be considerably more efficient since no international transmission circuits are used, resulting in lower costs.

CAMEL support is provisioned in the HLR on a per-subscriber basis. When a customer is marked for CAMEL service in the HLR, the switch-based SSP functions and SCP interactions can be invoked for all circuit-switched calls except emergency calls. When the SSP functions are invoked, events, such as call set-up, answer and disconnect can be reported to the SCP. The SCP can modify call handling at call set-up time, and can instruct the SSP to release a call at any stage. Forwarded calls are treated as mobile-originated calls.

The sets of detection points in the CAMEL Phase 1 and 2 standards are listed in Table 4.4.

### 4.5.3  CAMEL Phase 1

The CAMEL Phase 1 standard was finalised by ETSI in 1997 and it was part of the GSM Release '96 package, with conforming equipment becoming available around three years later.

The CAMEL Phase 1 call model has three trigger detection points (T_DPs) in the originating half of the basic call state machine (O_BCSM) and three in the terminating half of the basic call state machine (T_BCSM).

The gsmSSF in the MSC (or G-MSC) uses these trigger points to send requests for further instructions on call handling to the gsmSCF in an IN SCP. The T_DPs are statically armed, as with fixed-network IN procedures, but, unlike fixed-network IN, they are not pre-set in the SSP data – they are armed by virtue of the fact that they are provisioned in the HLR.

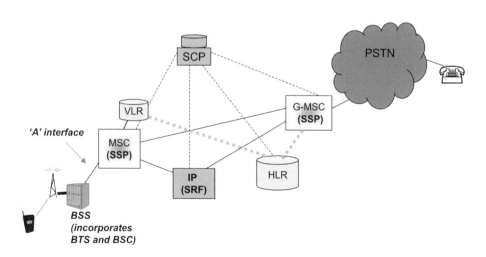

*Figure 4.8   GSM enhanced with CAMEL IN features*

*Table 4.4   CAMEL IN detection points*

| CS-1 (Q.1214) detection points (DPs) | CAMEL Phase 1 DPs | CAMEL Phase 2 DPs |
|---|:---:|:---:|
| **In the originating BCSM:** | | |
| 1 – Orig_Attempt_Authorised | | |
| 2 – Collected Info | ✔ | ✔ |
| 3 – Analysed Info | | ✔ |
| 4 – Route_Select_Failure | | ✔ |
| 5 – O_Called_Party_Busy | | ✔ |
| 6 – O_No_Answer | | ✔ |
| 7 – O_Answer | ✔ | ✔ |
| 8 – O_Mid_Call | | ✔ |
| 9 – O_Disconnect | ✔ | ✔ |
| 10 – O_Abandon | | ✔ |
| **In the terminating BCSM:** | | |
| 12 – Term_Attempt_Authorised | ✔ | ✔ |
| 13 – T_Called_Party_Busy | | ✔ |
| 14 – T_No_Answer | | ✔ |
| 15 – T_Answer | ✔ | ✔ |
| 16 – T_Mid_Call | | ✔ |
| 17 – T_Disconnect | ✔ | ✔ |
| 18 – T_Abandon | | ✔ |

When the gsmSCF receives a request from a gsmSSF, it is can respond in order to instruct the gsmSSF to do one of three things:

- connect the call to a specified destination
- continue with normal processing
- release the call.

The gsmSCF can also set Event Detection trigger Points (E_DPs) for 'answer' and 'disconnect'. E_DPs are dynamically armed by Request Report BCSM (RR_BCSM) instructions sent from the gsmSCF to the gsmSSF.

The SS7 IN signalling protocol for CAMEL is called the CAMEL Application Part (CAP), and is based on ETSI core INAP, but with a reduced set of operations. Whilst CAMEL Phase 1 has fewer INAP operations and triggers than CS-1, it is still a powerful capability. The key events in a call (call set-up, answer and disconnect) are included.

Finally, a new interface to the HLR, using additional and modified SS7 MAP operations, was provided in CAMEL Phase 1. This enabled the SCP to find out about mobile status, such as busy or idle, and mobile location information, such as VLR number or geographic co-ordinates of the serving cell base site.

The IN feature set of CAMEL Phase 1 is a modest sub-set of IN CS-1, and because many operators had already installed proprietary versions of INAP, take-up of CAMEL Phase 1 has been steady rather than swift. However, Phase 2 provides a big increase in the number of detection points supported.

## 4.5.4  CAMEL Phase 2

CAMEL Phase 2 [31] is backward compatible with Phase 1 and uses a more comprehensive BCSM, with a greater range of options for charging of Pre-pay calls. The INAP Apply Charging and Apply Charging Report operations were included in the enhanced CAP for Phase 2.

However, the main addition for Phase 2 is a set of procedures for including IN Intelligent Peripheral nodes (containing gsmSRF) in mobile calls. This enables the gsmSCP to control transactions in which users can buy more credit or make balance enquiries.

Also, Phase 2 extended the list of E_DPs. For instance it included the O_No_Answer E_DP, which allowed the CAMEL service environment (CSE) to provide a diversion on no reply service.

In summary, CAMEL Phase 2 provided the following extras on top of Phase 1:

- improved set of IN detection points and enhanced CAP signalling
- conditional E_DPs and additional E_DPs such as abandon, busy and no answer
- announcements, tones, voice prompting and information collection using an IN IP
- controls for interactions between network and SCP-based services
- new capabilities for tariff control (for Pre-pay services) and charging information control
- alerting pattern control.

## 4.5.5  Future CAMEL phases

Phases 3 and 4 of CAMEL address GPRS and SMS interworking with IN functions. Some IN CS-2 functions are also included to enable leg manipulation, e.g. for playing announcements during a call [47].

## 4.5.6  Wireless intelligent networking (WIN)

CAMEL, developed by ETSI as an extension to GSM, is the European regional standard for applying IN to mobile systems. The equivalent regional standard in the USA is WIN (wireless intelligent network). WIN was developed by the TR45.2 work group of the US standards authority, TIA (Telecommunications Industry Association), using ITU-T IN principles as a starting point. This was in acknowledgment of the widespread international success of the ETSI GSM standard and a desire to ease the future interworking between the two.

Whilst there are underlying differences in air interfaces and network signalling protocols, the WIN and CAMEL standards groups aimed to harmonise as closely as possible in order to ease international standardisation of the IN interface across GSM and non-GSM networks (e.g. ANSI-41 networks in the US).

CAMEL is based on the ITU-T Q.1214 CS-1 BCSM and WIN is based on the ITU-T Q.1204 [40] generic BCSM. Although the latter contains more PICs and DPs than the former, this common standards ancestry ensures a shared pedigree for the CAMEL and WIN standards whilst additionally providing WIN with linkage with the US AIN standard, which uses a similar call model to that in Q.1204.

## 4.6 ITU-T IN CS-2

The ITU-T significantly widened the scope of the IN standard for CS-2 in several directions, but the prime feature, and the one which received the most attention in terms of detailed definition, was 'call party handling' (CPH – otherwise known as 'leg control'). This is an enhancement to the call modelling principles of CS-1, and it enables us to use the IN to assist with the provision of multi-party calls, such as conference calls, call waiting and call transfer services. The CS-1 BCSM was principally concerned with influencing the sequential flow of a call as it progressed through the different set-up phases.

With different emphasis, CPH seeks to allow the remote service logic to manipulate the individual legs of a call. This is made possible because the originating and terminating call segments of CS-1 are now supplemented with a third 'associated' call segment, which makes it possible to model call party handling.

### 4.6.1 Extra functions provided in CS-2

The CPH mechanisms are discussed in further detail in Chapter 5, and the other extended functions specified for CS-2 [32] are summarised here.

*End-user access to call and service processing*. Greater attention is given in CS-2 to providing for IN service access for ISDN (primary and basic rate users) than was the case with CS-1.

*Service invocation and control*. As with IN CS-1, CS-2 limited its scope to single-ended, single-point-of-control services (called 'type A' services). Otherwise, IN-CS2 aimed to provide a 'super-set' of the IN-CS-1 capabilities, extending the range of service possibilities, but still predominantly attending to voice networks.

*Internetworking*. As illustrated later in the DFP diagram for CS-2 (Figure 4.10), support for data communication between networks is included, using SMF–SMF, SDF–SDF and SCF–SCF interfaces. The SCF–SCF interface recommendation is designed for the exchange of data or distribution of service logic and not for service scenarios involving distributed service control. The new mechanism will allow an SCF to query another SCF, or to hand over the call to another SCF.

*Out-channel call-related user interaction (OCCRUI)*. For communication between the user and the service control function in CS-1, an IP (intelligent peripheral) interaction was necessary to collect DTMF digits, or spoken responses, in answer to an audible menu choice. CS-2 defines a mechanism for exchanging service-independent information elements directly between the user's equipment and the SSF for onward transfer to the SCF. This feature makes it possible for a terminal to communicate directly with an SCP, for example to provide further address information, or respond to requests for choices of call processing options.

***Out-channel call-unrelated user interaction (OCUUI).*** This feature allows IN to assist in cases where non-call-associated features need to be invoked independently of call contexts. Examples are terminal registration, location updating and message waiting indicator transfer. OCUUI can therefore assist with convergence between fixed and mobile networks. It can enable SSPs to provide access to home location registers (HLRs) in the same way that GSM dedicated switches do. To model the required state changes for OCUUI, CS-2 introduces a new state transition model called the Basic Call Unrelated State Model (BCUSM).

The new functions introduced for OCUUI are the CUSF (call unrelated service function) and the SCUAF (service control user agent function). The BCUSM (in the CUSF) exists in parallel with the BCSM (in the SSF), located typically in a CS-2 SSP. Typically the SCUAF would be expected to reside in an ISDN CPE device, using D-channel signalling for network access. The CUSF contains the functions that allow the SCF and SCUAF to interact, detecting call-unrelated triggers and supporting the ensuing SCF transactions.

***Wireless access.*** IN CS-2 also provides other enhancements for the support of wireless and mobility systems. New functions for radio systems include:

◇ CCAF+ (call control agent function plus)
◇ CRACF (call-related radio access control function)
◇ CURACF (call-unrelated radio access control function)
◇ RCF (radio control function).

***Feature interactions.*** As the complexity of IN systems expands, the scope for adverse interactions amongst IN and non-IN features expands with compound interest. Q.1224 and Q.1222 lay out principles for the avoidance of interference between features and services, but the application of these principles is very much implementation-dependent and will vary across different network operators' IN environments. The handling of feature interactions, which is fundamental to successful IN implementations, has been discussed earlier (Chapter 2). The CS-2 proposed formal method for feature interactions is described below in the context of the IN CS-2 service plane.

***Extra trigger point.*** To supplement the CPH functions an additional BCSM PIC (O-suspended) and DP gave improvements in 'follow-on' call handling. This allows three-party calls (e.g. for call waiting) to be handled more satisfactorily than in CS-1.

***Service creation and service management.*** In addition to an extra six SIBs for new CS-2 functions, CS-2 also provided some added coverage for service creation and service management procedures, as well as for inter-SMF (service management function) communications. However, by comparison with the detailed text on CPH, these are high-level reviews. Because of the practical predominance of proprietary solutions in these areas, they have turned out to generate less attention in multi-vendor scenarios than the CPH network recommendations provided by CS-2.

The SMF is responsible for the control of the real-time IN entities, such as SSF, SCF and SRF, including service provisioning, data updates (to service data as well as customer-specific data), etc. The CS-2 SMF is based on the TMN functional architecture [50, 51]. Appendix B3 of Q.1224 explains the mapping between the service creation and service management functions for CS-2 and function blocks in the TMN functional architecture.

CS-2 also defines a service management access function (SMAF) for interfaces between SMF and the end-user and between SMF and the network or service operator.

The service creation environment function (SCEF) provides the toolkit for the preparation for installation of new IN services and for modifying existing ones. After the SCEF has prepared the new service (or altered an existing one), it dispatches it to the SMF (Figure 4.10), which then has responsibility for deploying the service. This responsibility includes version control, compatibility tracking, maintenance and fault handling aspects. The SMF also looks after real-time triggering issues, such as service key allocation, number plan management and SLP identification in the SCF.

### 4.6.2  Service plane for IN CS-2

Unlike CS-1, CS-2 aimed for a disciplined 'top-down' architecture approach, which would characterise and define the required services first and then work downwards through the INCM planes towards suitable implementation recommendations. This did not happen so emphatically with CS-1 because at that time the drive was to evolve existing network structures towards the concepts of IN. What could be done was highly dependent on what was possible in existing networks, so the approach was essentially 'bottom-up'.

As well as specifying this top-down process, the CS-2 service plane document [33] provided some initial steps towards a rigorous methodology for dealing with feature interactions using UFM (unified functional methodology). With UFM, all services (and eventually network architectures) are described using recommended functional descriptions and information flows. Basically, by carefully following the CS-2 'top-down' ideal, all services descriptions would use the same recognised methods, and so the identification of potential interactions between the features would become automatic.

### 4.6.3  Global functional plane for IN CS-2

New SIBs were included in CS-2, including some for CPH leg control operations as well as SIBS for starting, ending and controlling parallel SIB chain execution (the last four on the list in Table 4.5). The idea of parallel processing was new in CS-2,

*Table 4.5  The extra SIBs in IN CS-2*

- ☐  Join
- ☐  Split
- ☐  Create service process
- ☐  Send
- ☐  Wait
- ☐  End

allowing more than one SIB chain to be executed at once. To keep the execution of SIB chains in step with each other, points of synchronisation were also introduced. Two new SIB groupings were introduced, namely 'high level SIBs', which are groups of lower level SIBs, and 'service processes', which are groups of high level SIBs and lower level SIBs. These new groupings are used to simplify the logic descriptions when parallelism is used [34].

The global functional plane contains the 'global' (i.e. abstract) service logic (GSL), which defines the operation of the SIBs, and their interactions with the basic call process model to provide particular services. The GSL defines the point of initiation in the BCP where the logic of the specified SIBs is executed. The GSL also defines the point of return to the BCP.

### 4.6.4 CS-2 distributed functional plane

As with CS-1, the CS-2 DFP describes the CS-2 network functions independently of their physical locations. CS-2 represented a major expansion to the coverage of the ITU-T IN recommendation, and there were many candidate network features competing for inclusion. However, the functions that could be included in the CS-2 release had to achieve a balance between the CS-2 service expectations and the limitations imposed by the capability of the installed base of network technology to evolve with sufficient flexibility to make the upgrade feasible (Figure 4.9).

The actual DFP model for IN CS-2, showing the CS-2 functional entities and their relationships is reproduced in Figure 4.10. The five new FEs modelled in

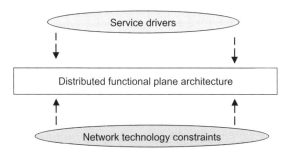

*Figure 4.9  The DFP scope constraints*

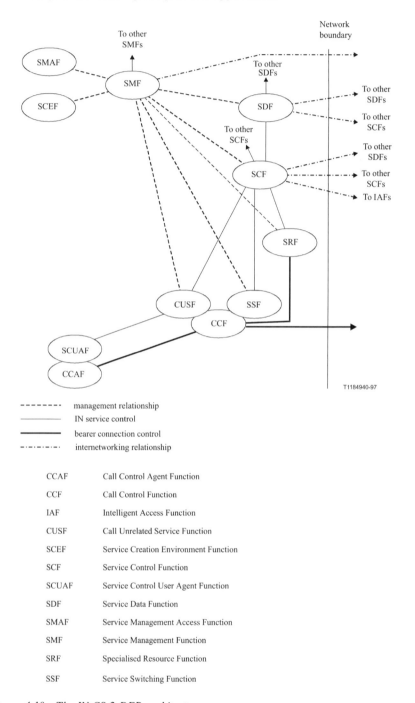

*Figure 4.10   The IN CS-2 DFP architecture*

Reproduced, with permission, from Figure 3-1/Q.1224 Fascicle 1/3 ITU-T publication, 'Distributed Functional Plane for Intelligent Network CS-2' [09/97].

CS-2, additional to those in the CS-1 DFP, are explained in Section 4.6.1 above. These are the SCUAF, CUSF, SMF, SMAF and the SCEF.

The CS-2 distributed functional plane recommendation [32] is a large document and mainly comprises detailed exposition of the functions and relationships represented in Figure 4.10. It explains the information flows, in terms of operations and parameters, and includes separate annexes for the coverage of mobility, TMN, network and fault management, and testing.

In addition to the new FEs introduced, Figure 4.10 differs from Figure 4.6 (the equivalent CS-1 diagram) in three other general respects:

1. Peer relationships now exist for SCFs, so that service logic can be distributed onto multiple SCFs using the new SCF-SCF interface. Similar peer relationships are provided for SDF and SMF functional entities.
2. Internetworking. As an extension to these new peer relationships, connection to other networks is now recognised as being standardisable. An 'intelligent access function' (IAF) is introduced for connections between IN and non-IN networks.
3. User to network signalling (known as 'out channel' signalling) is included. This uses the new SCUAF and CUSF functional entities.

### 4.6.5 IN CS-2 documents

The CS-2 recommendations were finalised in 1997. The 4-plane INCM, which provided the framework for the IN CS-1 document structure (Section 4.3.3 above), applies equally for the CS-2 recommendations. Therefore the numbering for the recommendation document series follows the same pattern as the IN CS-1 series, except that the third digit is a '2', for CS-2. Table 4.6 lists the CS-2 document set. The covering document is Q.1220, which explains the IN CS structures and includes the index for each Q122x document. Q.1290 contains terms and definitions.

### 4.6.6 Physical plane for IN CS-2

Q.1225 defines the CS-2 physical plane, which is reproduced in Figure 4.11. This is similar to Figure 4.7 (the CS-1 physical plane) except that it shows the extra

*Table 4.6   The ITU-T IN CS-2 documents*

| | |
|---|---|
| *Q.1220* | Q series recommendations structure for IN CS-2 |
| *Q.1221* | Introduction to IN CS-2 |
| *Q.1222* | Service plane for IN CS-2 |
| *Q.1223* | Global functional plane for CS-2 |
| *Q.1224* | Distributed functional plane for CS-2 |
| *Q.1225* | Physical plane for CS-2 |
| *Q.1228* | Interface recommendations for CS-2 |
| *Q.1229* | User guide for CS-2 |

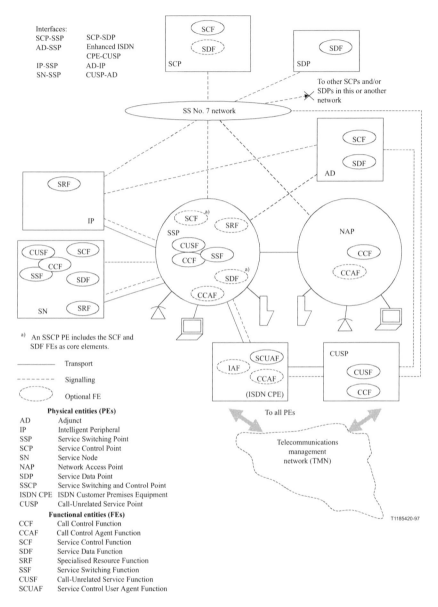

*Figure 4.11    Physical plane architecture for IN CS-2*

Reproduced, with permission, from Figure 1/Q.1225 ITU-T publication, 'Physical Plane for Intelligent Network CS-2' [09/97].

physical entities associated with the new IAF, TMN and call-unrelated functions included in IN CS-2.

## 4.7 ITU-T IN CS-3

In December 1999 a scoping document (Q.1231 – Introduction to IN CS-3) for IN CS-3 was produced, as well as a description of the IN CS-3 management information model. Six months later, the series of Q.1238 interface recommendations was released.

In general terms CS-3 consolidates many of the principles of CS-2, with some modest expansion in certain areas. For instance, there is some support, albeit limited, for broadband IN using the CS-2 call model. An IN-controlled number portability service is addressed, there are enhancements to the CS-2 CUSF (call-unrelated service function) and IN-ISDN interworking is improved. Multiple points of control are allowed, in the sense that CS-3 permits multiple triggering from the same SSF to different services, which can then run concurrently (this was permissible in CS-1 and CS-2 if the triggers occurred in separate SSFs). To allow for multiple points of control the FIM was enhanced to deal with ensuing interactions between the SSFs involved in the invocations.

### 4.7.1 IN CS-3 documents

The new documents for CS-3 are listed in Table 4.7. Where no new document appears, the CS-2 recommendations are carried forward for CS-3.

## 4.8 ITU-T IN CS-4

The CS-4 'introduction' [41] and 'distributed functional plane' [42] documents set out the scope of the topics that are expected to be central to IN CS-4

*Table 4.7   The ITU-T IN CS-3 documents*

| | |
|---|---|
| *Q.1231* | Introduction to Intelligent Network Capability Set 3 |
| *Q.1236* | Intelligent Network Capability Set 3 – Management Information Model Requirements and Methodology |
| *Q.1238* | Interface Recommendation for IN CS-3 |
| *Q.1238.1* | Interface Recommendation for IN CS-3 – Common Aspects |
| *Q.1238.2* | Interface Recommendation for IN CS-3: SCF-SSF Interface |
| *Q.1238.3* | Interface Recommendation for IN CS-3: SCF-SRF Interface |
| *Q.1238.4* | Interface Recommendation for IN CS-3: SCF-SDF Interface |
| *Q.1238.5* | Interface Recommendation for IN CS-3: SDF-SDF Interface |
| *Q.1238.6* | Interface Recommendation for IN CS-3: SCF-SCF Interface |
| *Q.1238.7* | Interface Recommendation for IN CS-3: SCF-CUSF Interface |

implementations. Many new service features and corresponding network capabilities [46] are outlined; however, the prime focus of CS-4 is the establishment of IN's role against a background of converging IP (Internet protocol) and PSTN networks. The scope of CS-4 includes support for IN access from H.323 Gatekeepers and SIP Proxy Servers, although this support is limited to the BCSM approach, and does not to date extend to the CS-2 call party handling functions.

### 4.8.1 Alliance between IN and Internet telephony

The CS-4 documents available at the time of writing demonstrate an increasing desire to align with the Internet standards coming from the IETF. In particular, CS-4 acknowledges and builds on the Internet-driven initiatives such as PINT [43] and SPIRITS [44].

**PINT** ('PSTN/Internet interworking') transactions are characterised by requests that originate via an IP network and result in the automatic invocation of telephone calls. An example could be an Internet user clicking on an icon on a travel agent's Web page with the intention of eliciting a return telephone call from the travel agent giving more information.

The interface from an IP server to a PSTN could be through a PBX, an ISDN device or a mobile phone network. However, the main IN interest is the connection of an SCF (through the IAF) to a PINT gateway. This then allows calls to be set up automatically across the PSTN, using IN network features, on receipt of PINT requests.

**SPIRITS** ('services in the PSTN/IN requesting Internet services') is concerned with requests in the other direction – from the IN into an IP network. The SPIRITS IP client could be built into an SCF, located in an SCP or SN, for instance, and originate requests to a remote IP server at significant stages during a PSTN call. Internet call waiting (where a user whose only phone line is engaged on an Internet connection is alerted to the arrival of an incoming call attempt to the phone) is a typical SPIRITS application.

In another example of the use of SPIRITS technology, a corporate Freephone customer might subscribe to a service that provides statistics at an IP server on failed incoming calls. In this case the IP server could be alerted by a SPIRITS transaction that is invoked whenever a 'routeset fail' E_DP (event detection point) is triggered in an incoming call to that customer.

### 4.8.2 Interworking between CS-4 and other networks

The emphasis of CS-4 is on interworking with surrounding networks. CS-4 expands on the use of the IAF (intelligent access function), which was introduced in CS-2, in order to interwork with non-IN structured networks. As outlined above, it also includes preparatory information for interworking with SIP, H.323, PINT and SPIRITS systems.

Specifically CS-4 also defines gateway functions for service applications and service control. These are:

◇ SA-GF (service application gateway function)
◇ SC-GF (service control gateway function).

*4.8.3 Service application gateway function*

The SA-GF provides access to distributed service logic via an open API, including the necessary firewall and security mechanisms. Typical API technologies are CORBA, JAVA and JAIN. Figure 4.12 shows SA-GF as embedded in an SCF, which is then able to provide the open API with access to the PSTN via INAP interfaces to SSFs and SRFs, etc. However, another CS-4 option is for a new interface to be defined between separated SCFs and SA-GFs. ITU-T CS-4 is also linked into the OSA API definition [37], which is introduced in Chapter 6 in connection with the Parlay API evolution.

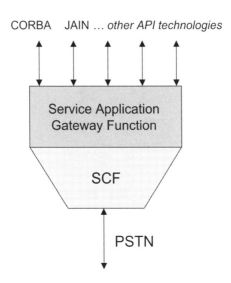

*Figure 4.12   Service application gateway function*

### 4.8.4  *Service control gateway function*

The SC-GF allows access to existing IN service logic from an IP[13] network node. It could be co-located in either the SCF or IP (Internet protocol) server, or separately implemented, as shown in the interworking example of Figure 4.13. This diagram illustrates an IP network call control executive on the right-hand side, accessing traditional IN functions on the left-hand side. This call control executive in the IP network is typically an SIP proxy, an H.323 gatekeeper or a media gateway controller (MGC), so these are the logical VoIP components that would be modelled by an IN SSF as shown.

The SC-GF provides the necessary lower-level mapping functions for INAP messages between the two networks.

At the call and bearer control level, speech trunks terminate on a media gateway and the ISUP signalling terminates on a signalling gateway, which then interacts with the IP control element. Because the ISUP signalling is no longer associated

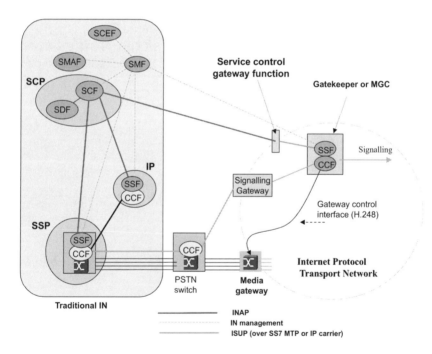

*Figure 4.13　Service control gateway function – an example scenario*

[13] At this stage of the book there is scope for confusion with the abbreviation 'IP', which is used for both the IN 'intelligent peripheral' and 'Internet protocol'. For instance, in Figure 4.13 both terms appear. We therefore take care to clarify usage each time, but where the term 'IP network' is used this invariably refers to Internet protocol.

with the bearers that terminate on the media gateway, a channel control interface is required between the IP call control entity and the media gateway. ITU-T H.248 (based on Q.931 and also known as MEGACO) is the international standard interface for this. For further information on VoIP, including the MEGACO, H.323, SIP and BICC aspects, the reader is referred to Reference 52.

The different IN functional entities, including the 'new' SSF in the IP network, are managed by the IN SMF. This includes, for instance, the population and update of locally held IN data such as the SSF trigger criteria.[14]

### 4.8.5  CS-4 service features

Q.1241 describes several new service features, mostly drawing on the recent convergence work between IN and Internet protocol technology. Q.1244 [42] provides descriptive examples of how such services can be provided using PINT, SPIRITS and SIP in conjunction with traditional IN network entities. The new service features include:

◇ **End-User Service Data Customisation** (this is via an IP network, using PINT access).
◇ **Request-to-Call-back** (an SCP sets up a call, instigated by a user clicking on a Web page icon).
◇ **Request-to-Call-back via IP** (in this case the call proceeds as a VoIP call).
◇ **Internet Call Waiting**.
◇ **Web-controlled PSTN/IP Conferencing**.

### 4.8.6  IN CS-4 documents

The following recommendations (Table 4.8) apply specifically to CS-4. The service plane, GFP and physical plane have not changed, so in these cases the previous CS-2 or CS-3 documents still apply to CS-4.

*Table 4.8   The ITU-T IN CS-4 documents*

| | |
|---|---|
| *Q.1240** | IN CS-4 Recommendation Structure |
| *Q.1241** | Introduction to IN CS-4 |
| *Q.1244** | Distributed Functional Plane for CS-4 |
| *Q.1248** | Interface Recommendation for IN CS-4 – Common Aspects |
| *Q.1248.2** | Interface Recommendation for IN CS-4: SCF-SSF Interface |
| *Q.1248.3** | Interface Recommendation for IN CS-4: SCF-SRF Interface |
| *Q.1248.4** | Interface Recommendation for IN CS-4: SCF-SDF Interface |

\* *Approved pre-published versions of new recommendations, July 2001.*

---

[14] For diagrammatic clarity the speech circuits between SSP and IP are not shown in Figure 4.13. The SSP–IP connection is shown as ISUP – it could equally well have been Q.931 (Section 2.4 refers).

# Chapter 5
# Call party handling

*Airplanes are interesting toys but of no military value.*
  Marechal Ferdinand Foch, Professor of Strategy, Ecole Superieure de Guerre.

## 5.1 Introduction

The ITU-T CS-1 IN standard, along with the subsequent ETSI core INAP modifications, was a resounding success. It led to IN installations in most major operators' networks because the equipment vendors generally supported the engineering content of the recommendations, or at least a significant sub-set of them. However, in the period leading up to the original CS-1 publication there had been lively debates about the basic network model that should be chosen for the standard. The contenders were the sequential call logic approach, which later became known as the 'basic call state model', and the connection view state (CVS) model, based on 'snapshot' views of call-states including the connection configurations. The CVS model included transition rules that prescribed the possible connection states between the 'legs' that represented the end-points. In the end the CS-1 standard makers settled on the basic call state model, which cycles through the logical sequence of events on the originating and terminating side, and this approach has generally been the subject of the earlier chapters of this book.

Essentially the debate was about how to represent the mechanics of the underlying switch network to the IN service logic in the SCF. The SCF does not need to be aware of too much of the detail of the telephone network; on the other hand it needs a logically consistent view, comprehensible by the SSF as well as the SCF, of the service logic's context. This representation is known as the 'call model', as we have discussed earlier.

The debates over the nature of the CS-1 call model became complex and tended towards the philosophical, but the upshot was that the BCSM approach was chosen for CS-1. The IN switching manager, using the CVS approach, was deep-seated in the original ITU-IN framework document series. It was for instance described in overview terms in Q.1204, which is the IN generic distributed functional plane architecture and it was introduced into the CS-1 recommendations [9]. However, it was not fully adopted as a standard recommendation until IN CS-2 in 1997.

Whilst CS-2 is a preferred basis for the development of IN services, in the mid-1990s the major European network operators could not wait for the culmination of the lengthy standards process that was in place for CS-2. They therefore seized on the newly available and vendor-endorsed ETSI Core INAP, and invested in the necessary switch upgrades and in service control equipment. In practice the network upgrades needed to achieve an IN architecture usually had to be justified step-by-step in discrete business cases, each being justified by revenue-projections of the services that could be realised at each step. The introduction of CS-1 was therefore a staged affair, often taking several years. Hence, when the CS-2 standard was finally agreed, and the agreed CVS model ratified as an international standard, it was too late. The established network operators had, for the most part, embarked on expensive CS-1 based programmes and another fundamental change to the basic call model so quickly after the first one was unrealistic and could not be justified. Also, most of the business case arguments that would have applied to IN CS-2 had in fact already been used earlier to justify the implementation of CS-1 upgrades to the existing switches!

However, new operators, without a legacy of investment in CS-1, are well placed to take advantage of the more powerful service control features that IN CS-2 offers. Otherwise, for a network operator with an existing set of legacy IN CS-1 services, it becomes necessary for the IN versions to co-exist when the need arises for an upgrade to IN CS-2 leg control capabilities. Some or all of the network's SSPs need to be enhanced with the new call model, and CS-2 enabled SCPs must be introduced. A typical scenario might be as shown in Figure 5.1, where an SCP that

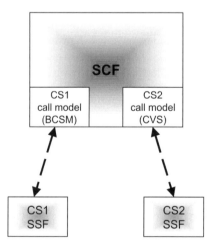

*Figure 5.1   Coexisting CS-1 and CS-2 call models*

can handle both interface types is able to view dependent SSPs either directly through the CS-1 BCSM or via the CVS modelling structures. As we shall see shortly, the CVS approach also includes the BCSM, but at a more detailed level. In the interests of clarity, Figure 5.1 does not show the detail of BCSM on the CS-2 call model side of the SCF.

### 5.1.1 Early 'leg control' network functions

In order to more substantially introduce the concept of call 'legs' before we go on to explain details of call party handling mechanics, it is probably helpful to review some of the familiar CS-1 INAP operations, which we are familiar with from Chapter 2, in terms of their 'leg-control' aspects. Several of these earlier operations implicitly involve the notion of call legs, as we now demonstrate:

- The **Connect** operation requests the SSF to attempt to connect a waiting A leg (representing a caller's line or an incoming speech circuit in an inter-exchange trunk group) to a new B leg. The B-leg's local identity is obtained by the SSP by analysis of the destination address that is provided as a parameter of the Connect operation from the SCF.
- The **ConnectToResource (CTR)** operation connects a waiting A Leg to a B leg that represents a channel in a trunk group leading toward a physical entity – typically an IN intelligent peripheral containing an SRF module. CTR is also commonly used for extending a call to an announcement resource on an internal channel within the SSP. In this case the B leg represents the internal channel connection to the built-in hardware.
- The **DisconnectForwardConnection (DFC)** operation is sent from the SCF to request the SSF to disconnect a B leg that had previously been set up (e.g. in a CTR or ETC operation) to the channel resource hosting an SRF module.
- The **EstablishTemporaryConnection (ETC)** operation is sent from the SCF to request the SSF to set up a temporary connection between the calling party's A leg and a B leg representing a connection to another network node, such as an IP, or another SSP, hosting the target SRF module.
- The **InitiateCallAttempt (ICA)** operation is sent from the SCF to request the SSF to create a new call to a call party identified by address information provided by the SCF in the associated DestinationRoutingAddress parameter. The new call leg can then be connected to an exchange announcement. This would be carried out for instance by using the CTR and PA (PlayAnnouncement) operations. However, in CS-1 this operation cannot normally be used (as it can be in CS-2) to connect together two network destinations.
- The **ReleaseCallPartyConnection** operation is sent from the SCF to request the SSF to release a specified leg of a call.
- The **RequestReportBCSM** operation requests the SSF to monitor (and report on) certain call-related events. It does this by arming event detection points. The leg to be monitored is specified within the operation's parameters.

*5.2 The CS-2 call model*

The main components of the CS-2 call modelling activities in the SSP are shown in Figure 5.2.

Towards the bottom end, the basic call manager (BCM) acts on the bearer control function and is an abstraction of the fundamental call control functions and event detection mechanisms in the switch. The BCM contains the BCSM, including the detection points and trigger criteria needed for IN invocations. In a similar manner to CS-1 (described in Chapter 2), the BCSM identifies the sequential sets of call processing events that have relevance for external service control. The CS-2 BCSM is an expansion of the CS-1 BCSM within the context of the overall ITU-T BCSM of the ITU-T IN generic framework distributed functional plane of Q.1204.

The feature interaction manager (FIM) is shown above the BCM in Figure 5.2, and this performs similar functions to those described for CS-1 earlier, in Chapter 2.

At the upper level of the SSP model in Figure 5.2 is the IN switching manager, which is responsible for providing the 'aperture' through which the SCF views the call control activities of the switch. The IN switching manager provides the

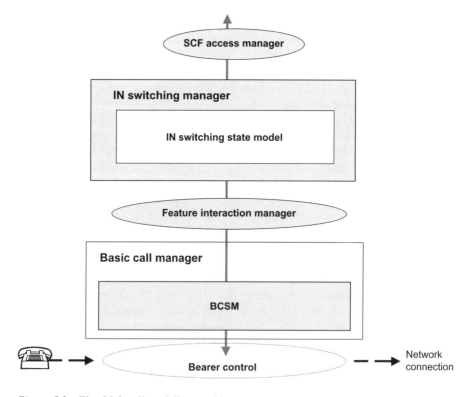

*Figure 5.2   The CS-2 call modelling architecture*

controlled access for external service logic to act on the lower level switch resources of the SSP. The operating service logic for the switching manager is located mainly in the IN switching state model (SSM).

As we mentioned earlier, the overall shape of the SSP call model in Figure 5.2 is similar to that originally laid out for CS-1. However, as we show in the following paragraphs, CS-2 expands on the detailed contents of the IN SSM, which is effectively the 'engine room' of the switching manager. The IN switching manager had already been outlined in the international standards before the publication of the CS-2 recommendations in the context of the scope of the overall ITU-T IN framework in the CS-1 recommendation. With this background, CS-2 provides the essential operational detail required in order to lead into practical implementations of the CPH principles.

## 5.2.1 Legs and connection points

The term 'call party handling' (CPH) is fundamental to the CS-2 connection view approach that we are describing here. CPH refers to a system of modelling telephone calls where call 'legs' (representing bearer channels or end-points) can be added to, joined into or deleted from a connection point. The advantage of this approach is that the manipulations required for telephony services such as 'call transfer', 'call hold' and 'multi-party calling' can be fairly easily represented to the service logic. The sequential call flow model of CS-1 can, by comparison, be cumbersome in its attempts to model these types of supplementary services.

For CS-1, the only view seen by the SCF was the progress of the call as it stepped through the different 'points in call', and this could be seen at both originating and terminating sides. This meant, however, that SCF manipulations were more or less limited to calls involving just two parties, unless multi-leg involvement was modelled through sequential actions. However, this is not perfect because of the intervening time delays. A service example (a 'call completion for directory enquiries' service feature) that illustrates the way in which CS-1 has to emulate multi-party connections with sequential state-machine operations is discussed in Chapter 7.

With the CPH approach, the service logic has an additional viewpoint, and it now becomes aware of a call as a whole, rather than in halves. A consequence of this approach is that the addition of more 'legs', or call parties, for multi-party calls becomes easier to model. CPH allows legs representing bearer channels to be added, deleted, joined and separated from the other legs of a call using the functions introduced in CS-2.

In CPH, CS-2 uses object-oriented techniques to describe the IN SSM and the underlying CVS concept. The CVS model gives an overview representation of the state of the underlying calls and their connections in terms of connection view 'objects'. The connection view objects used in the model are:

- call segments

- connection points
- call segment associations
- basic call state machine (BCSM) instances.

Whereas a CS-1 call generally comprises one originating and one terminating BCSM instance, CPH in CS-2 can now include multiple (more than 2) BCSM instances in a call. This is why the BCSM instance is included in the above list of basic objects for call modelling scenarios.

Each BCSM instance used in particular modelling situations then has devolved responsibility for the recognition of IN-significant events (and their notification to the SCF) for a specified call segment. The notion of 'call segments' is explained further on in this chapter.

With the enhanced CS-2 call model, the SCF service logic now needs to be aware of the connection-state viewpoint as well as the (possibly several) BCSM instances for the legs involved in a call using IN features. The SCF therefore has to be aware of substantially more of the complexities of the underlying switch network than was necessary with a CS-1 BCSM-only approach. An indication of the consequent potential for increased implementation complexity is given by the relative number count for ITU-T CS-2 INAP operations. CS-2 specifies 74 SSF-SCF INAP operations, which compares with 26 for the CS-1R standard recommendation!

## 5.2.2 CS-2 core switch capabilities

To provide practical guidance for upgrading existing switches or SSPs to host the new call party handling functions, the ITU-T CS-2 recommendations [32] defined four basic capabilities. These are briefly summarised as follows:

1. Either of the parties involved in the call should be able to summon the attention of the service logic during the speech phase. This is known as a mid-call trigger detection, which could be expedited via a switch-hook 'flash' or the dialling of a nominated digit string. This might for example be used in order to arrange for a third party to be brought into the call.
2. It should be possible for any party in a call to be temporarily disconnected from an established connection and, temporarily or otherwise, be reconnected to a speech path resource, such as an announcement machine, under the control of IN service logic in an SCF. The switch should then be able to correlate the consequent transport connections, so these can be individually manipulated as required.
3. The IN switching manager can present the SCF with an abstract view of an isolated portion of a call, and this portion is a call segment, or a 'half-call'. The term 'call segment' refers to the collection of legs and connection points that represent the physical transmission and connection resources and the call processes (the BCSMs) that constitute that isolated portion of the call. This

concept of restricting the SCF's view to an abstraction of an isolated portion of a call is illustrated in Figure 5.3.[15]

4. Lastly, the switch should be able to combine, on instruction from the SCF, selected transport paths that have been independently manipulated or transferred, into a call connection. Figure 5.4 shows a case where an SCF is able to act on an association of two call segments. In this illustration user A is part of an existing call with user B and a new call attempt from user C is being set up towards user A. The controlling SCF is now able to see the developing association between a new call segment for the incoming third party call arriving at A's termination in conjunction with the established call segment representing the original outbound call setup towards user B.

### 5.2.2.1 The hybrid approach to call party handling

In order to provide an interim means of introducing some CPH features without heavy switch-upgrade overheads, a 'hybrid approach' using CVS principles was included in the CS-2 distributed functional plane [32] description.

We do not enter into the details of the hybrid approach here, apart from to say that it provides a means of allowing IN CS-2 services to be provided without a full CPH upgrade to the switches. This is achieved by using a bridging function in a separate and dedicated IP network node that contains the required CPH capabilities. This enables an expedient solution to be introduced while a reduced set of connection view states exists in the SSP. Details of the network options offered by the hybrid CPH solution are provided in Reference 32.

### 5.2.3 The connection view model states

We now look at the various connection view states that are possible in IN CS-2 CPH.

The enhanced CPH view of the SCF, which must now be aware of the connection state of the relevant call segment as well as keeping track of any BCSM models with which it has a relationship, is shown in Figure 5.5. As we have already mentioned, this is referred to as the 'connection view' model.

Connection view states (CVSs) describe the connection configurations for particular connection views. In order to illustrate the principles, Figure 5.6 shows the progression of a call segment from the 'null' state to the stable 2-party state for a simple 2-party call. This progression is achieved either through an originating call set-up or by the arrival of an incoming call attempt.

---

[15] This emphasises the fundamental difference, which we introduced earlier, between the new CS-2 call party handling and the CS-1 approach. With CS-1, the SCF was directly aware of the BCSM state of the call through the exchange of the CS-1 INAP operations. In the case of the expanded set of CS-2 INAP operations, the SCF is additionally able to see the state of the individual call segments in terms of their connection states.

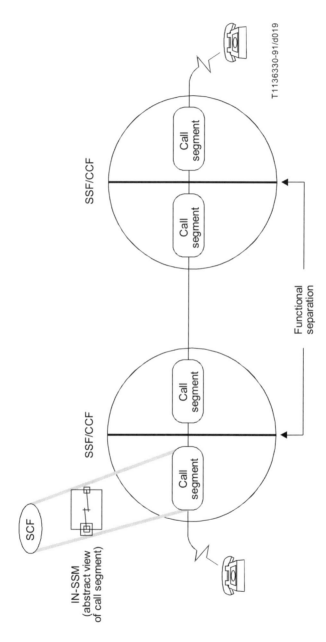

*Figure 5.3   Call segments in a two-party inter SSF/CCF call*

Reproduced, with permission, from Figure 4–13/Q.1214 ITU-T publication, 'Distributed Functional Plane for Intelligent Network CS-1' [10/95].

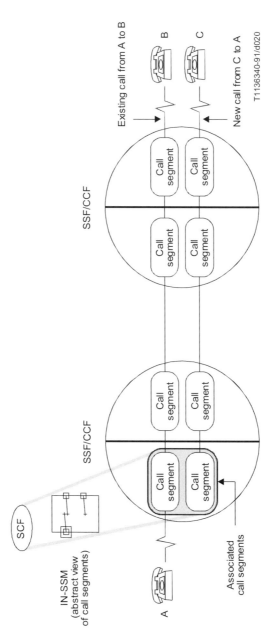

*Figure 5.4 Multi-party calls and associated call segments*

Reproduced, with permission, from Figure 4–14/Q.1214 ITU-T publication, 'Distributed Functional Plane for Intelligent Network CS-1' [10/95].

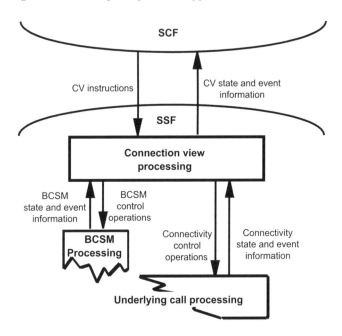

*Figure 5.5   The connection view model*

Reproduced, with permission, from Figure 4–18/Q.1224 ITU-T publication, 'Distributed Functional Plane for Intelligent Network CS-1' [09/97].

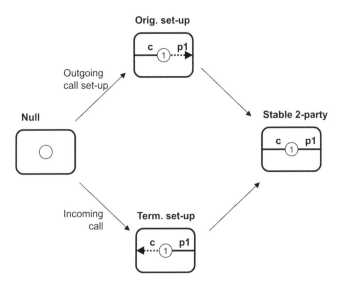

*Figure 5.6   Simple connection view states*

Figure 5.6 shows four CVSs, each with a single call segment, identified by the number in the central circle, which represents the connection point. Call segments are conventionally drawn as boxes with rounded corners. Legs are depicted by solid or broken lines, and there is always just one controlling leg (labelled 'c' in the CVS and always on the left hand side of the connection point) and relates to the originating or terminating access. On the right hand side of the connection point is shown the passive leg (or legs in the more complex call segments shown later). These are labelled 'p1', 'p2', etc.

A leg is shown as either solid or broken. We have also shown arrowheads on some of the lines to indicate call set-up direction. A solid line indicates 'joined' status, which means that the user attached to that leg is in communication with other users through the segment. A broken line indicates that the leg status is 'pending', as is the case with the 'Orig. set-up' and 'Term set-up' CVSs in Figure 5.6, or is 'shared' or 'surrogate'. 'Shared' means that a call segment is part of a CSA, and the user is associated with the controlling leg of the associated segment. An example is in the 'M-party set-up' CVS in Figure 5.7, where the controlling leg is shared between the two call segments. A controlling leg can be a 'surrogate' when it represents a user who is not actively participating in a call, as in the case of the 'transfer' CVS of Figure 5.7, where the controlling user has actually left the call to the two passive legs.

Having introduced the CVS concepts for the simple 2-party transitions of Figure 5.6, we now illustrate the power of the IN CS-2 CPH features by showing, in Figure 5.7, some of the more complex CVS configurations.

Figure 5.7 shows most of the CS-2 call configurations that are possible. They are labelled CC0 to CC12 (with some gaps). The diagram also shows several possible transitions, representing potential IN service features for CS-2. For instance, it shows how the 'stable 2-party' CVS of Figure 5.6 can progress to a multi-party call via the 'M-party set-up' and 'call on hold' CVSs. These last two CVS configurations each consist of two call segments and they are therefore CSAs (call segment associations).

There is obviously a finite number of ways in which call parties can be connected to each other, from the viewpoint of the SCF. IN CS-2, Reference 32 provides a catalogue of 14 example CVSs, and each one is unambiguously described in terms of its entry and exit events, and its relationship to the BCSM. Rules are specified, using object-oriented techniques, in order to describe the allowable transitions between the different connection view states. The resulting set of rules comprises the IN SSM – introduced earlier in this chapter (illustrated in Figure 5.2).

The documented [32] CVS examples are listed, with brief summary explanations, in Table 5.1.

### 5.2.4 Information flows supporting CPH

We now list, and briefly describe, the protocol information flows that are defined for the SSF–SCF interface to enable the call party handling functions, which we

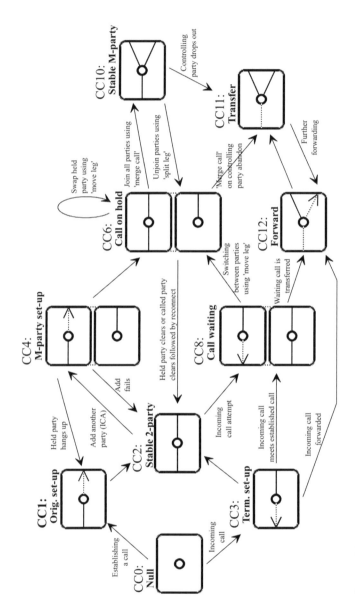

*Figure 5.7   More complex CVSs*

*Table 5.1   The connection view states*

| CVS | Explanation |
|---|---|
| Null | Call processing is inactive and no legs are connected to the connection point. |
| Originating set-up | A two-party call is represented in the set-up phase. |
| Stable two-party | A stable two-party call – can be originating or terminating from the controlling user's perspective. |
| Terminating set-up | A terminating two-party call in the set-up phase. |
| M-party set-up | The controlling party has put one (or more) party on hold while originating a new call to another party. Two (or more) associated call segments are involved. |
| Call on hold | The call has two associated segments. The controlling user has one party on hold whilst participating in a stable two-party call with a third party. |
| Call waiting | A stable two-party call is in progress and a new call attempt is targeted at the controlling user. |
| Stable M-party | A multi-party call in the active (speech) phase. |
| Transfer | The controlling leg has now disconnected, and has become a 'surrogate', whilst a stable call exists between the two passive legs. |
| Forward | The call is forwarded by the controlling party, who becomes the surrogate leg, and set-up is being established between the passive legs. |
| Originating set-up M-party | The controlling user is connected to an SRF, which has instigated a new call, which is in the set-up phase. |
| Active M-party | The call being set up in the 'originating set-up M-party' CVS has now reached the stable state. |
| One-party set-up | A single party call is originated by the network and is in the set-up phase. |
| Stable one-party | The 'one-party set-up' has now moved to the stable phase. |

have been describing, to be brought into play. Some of these operations are illustrated against the example CPH transitions that were shown in Figure 5.7 earlier.

The comprehensive list of the extra INAP operations for CS-2, and details on data values, are in Reference 32. The operations here are ordered alphabetically:

**CreateCallSegmentAssociation** – creates a new CSA. The 'Move Call Segment' operation (described below) can then be used to populate the new CSA.

**DisconnectLeg** – enables the SCF to release a specified leg from a call connection, leaving the remaining legs, and any connection between them, intact.

**HoldCallInNetwork** – allows a call to be queued in the set-up phase (e.g. to provide a 'call-completion to busy' service the call would be queued until the destination became free). The SCF can, in an associated parameter, provide information on what announcement or tone is to be played to the caller in the meantime.

**InitiateCallAttempt** – was specified earlier in CS-1, and it was intended for services such as network-originated alarm calls. At first sight its potential might have been be overestimated. For instance in CSI it was not normally used to originate a 'call' between two external parties. However, its flexibility improved with CS-2, where it became capable of being used to connect two external destinations together in a call.

**MergeCallSegments** – will merge two call segments which are 'associated' (i.e. they have a single controlling leg) into a single call segment. It can also be used to re-establish the bearer connection between two legs within one single 2-party call segment. We can therefore initiate a call (ICA) to an operator, whilst a caller is listening to a voice announcement, in preparation for through-connection. This could also be used for a 'pop-up' operator facility.

**MoveCallSegments** – moves a call segment from the source CSA to a target association. The controlling leg party must be the same, so it can have the effect of merging two associated call segments into a single call segment. However, this only applies during a terminating BCSM for a 2-party call segment, where the call processing has been suspended at T-busy T_DP. For this reason, a natural service application would be for an incoming call to a busy destination to be joined into the connection, after the call had been offered to (and accepted by) the target (busy) party.

**MoveLeg** – issued by the SCF to move a leg from one CS to another with which it is associated.

**Reconnect** – is used to re-establish communication between the controlling leg and the (held) passive leg(s) of a call with two or more parties, when the controlling leg has disconnected.

**ReleaseCall** – can be used for forward clearing of a B-party (after encountering a 'Disconnect' detection point, for instance).

**SplitLeg** – used to separate one call party from its CS. In the case of a multi-party CS, SplitLeg will place it in a new associated CS. It interrupts the speech connection between the leg to be split and the legs remaining in the original CS. This information flow is the reverse of MoveCallSegments (above).

### 5.3 Call party handling – A renewal of IN?

In many ways CPH in ITU-T IN CS-2 restores some of the 'intelligence' which was available in the network 50 years ago when the 'intelligent' component of networks was the large inter-connected web of manual-board telephone operators (see Figure 5.8). This network represented a distributed intelligence system that communicated on a separate plane from the bearer connections. The bearer connections were only switched through when the necessary transactions had been carried out to ensure that an end-to-end connection was available.

Service logic and data were available wherever it was needed for a wide range of services. Switchboard operators in those days could manipulate the legs involved in a connection. They could provide diversion services, they could put subscribers on hold whilst they offered the call to other subscribers, they could connect several subscribers together at once, and they could split a multi-party call into individual

*Figure 5.8   An early implementation of call party handling*

calls. The versatility of the 'sleeve control' [2] manual board system allowed the operator to provide many IN services, such as connecting several parties together at once or transferring a call from one party to another. The operator could then drop out of the call speech path altogether and return to monitor and supervise the call at any time. In fact they could do all the operations such as split, join, delete and separate that were subsequently specified in the IN CS-2 recommendations, in addition to providing most of the regular CS-1 services such as diversion on no reply and time-of-day routing to Freephone numbers.

*Chapter 6*
# Distributed intelligence

*For the strength of the Pack is the Wolf, and the strength of the Wolf is the Pack.*
Rudyard Kipling (1865–1936)

Up to this point in the book the controlling service logic has been presented as being located (at least logically) at a single point and this has been represented by the IN SCP. The concept of remote data has been introduced with the IN SDP, but it has generally been assumed that the controlling logic that runs the service is hosted by the network operator on a single computing facility. Thus we have service-switching points, and service control points, such that an SSP might communicate with a 'home', or pre-assigned, SCP for its service requests. For many services, such as those described earlier, this model has proved to be sufficient. Early intelligent networking was to do with a set of discrete database nodes serving large numbers of telephone exchanges, with a reliable and intelligent signalling system in between. This ensured that the required message reached the intended destination database processor, and that the reply would be correctly returned to the originating network node.

We now look at architectures where the intelligence is no longer necessarily fixed in a single location but is allowed to be distributed across a network. The use of flexible distribution of computing functions across processor clusters, and the consequent dynamic location of service logic within a single service control point is not new. However, extending this distribution of remote intelligence across a network comprising different vendors' equipment is still, in practice, relatively new ground for IN implementations, although theoretical modelling of the associated issues has been wide-ranging.

Broadly, there are two categories of argument in favour of distributing the intelligence in a network. Firstly, there are the practical operational reasons for distribution that are to do with improving network efficiency, and secondly there are the strategic drivers for open interfaces between network domains and service provider domains. We now discuss these categories separately.

## 6.1 Distribution of intelligence for network efficiency

Initially IN database processors tended to be large. They were points of focus for heavy real-time traffic demand. As we discussed in Chapter 2, the database computers themselves therefore needed to be highly reliable and fault-tolerant. This was because the IN translation data held there was critical for successful completion of a large proportion of the calls handled by the entire network, representing large amounts of revenue. Because of this criticality, local node resilience mechanisms were usually also provided to ensure that back-up processing and alternative call handling measures were available as expedients in case of network or equipment malfunction.

Because of the high cost of these early large fault-tolerant and highly secure processors, an alternative architectural approach has now become popular. This involves distributing the service control functions across larger numbers of general purpose, cheaper (but less resilient) servers. The required reliability levels are achieved by using redundancy-sparing techniques. This allows significant cost savings to be made because a cluster of general purpose Unix workstation-type platforms can be provided more cheaply than the earlier high-grade transaction processors. Similarly, at the network level, greater versatility becomes possible by distributing the service control function across a larger number of cheaper control points. This means that, as well as being able to 'freeze out' malfunctioning processors, or processor nodes, the network management systems are able to move traffic automatically from one part of a network to another as traffic patterns change through the day. At times of high load at one part of the network, IN service requests can be routed to alternative host computers by altering the SCCP global title translation data tables (explained in Chapter 3). Similarly if one processor cluster requires to be taken off-line for maintenance work, its load can be reallocated to peer processing nodes beforehand.

It has to be remembered, however, that a disadvantage of using distributed intelligence for increased flexibility and hardware cost-savings in this manner is that the supporting data management update mechanisms (as well as the SCCP global title routing management systems) obviously become more complex. This is because up-to-date data for IN translations must be made available wherever it is needed by the service control logic. Up-to-date versions of frequently used data may need to be held locally at all SCP sites, but it is obviously not efficient to distribute local copies of rarely used data to every control point in a large network just in case it is needed. In these cases it is more efficient for the requesting service logic to obtain the data from the source database when it is needed.

For example, a universal mobility service requires personal data profiles to be provided for each user. These customer profiles hold details on current location addresses, time-of-day routing plans, screening lists, etc. When a call attempt is made to one of these customers, the caller's local SSP will normally originate an IN trigger to a controlling SCP, which is unlikely to also be the host for the called customer's personal profile.

Therefore, either all customer profiles need to be distributed and updated at all points which could receive IN requests for this service, or a data request facility[16] is implemented so that local SCFs can query remote SDFs where the required data is hosted. In a large majority of cases the latter option is likely to be the most sensible, although some local data caching for heavily used destinations may be employed.

Another aspect of distributed intelligence for mobility services is illustrated in the CAMEL call flow examples of Chapter 7. In this case the controlling SSP makes an IN request to the home SCP for a roaming customer in order to access the required data.

## 6.2 Distribution of intelligence across open network interfaces

Since the late 1990s there have been strong commercial and legislative pressures for opening up public network infrastructures for external service providers. The reasoning for this is that the natural consequence would be an expansion of the telecommunications industry. Potentially lucrative markets have been identified for service provision companies (sometimes called 'soft-telcos'), whose main business is the provision of new services for their clients through existing public networks. These independent service providers do not invest in their own information transport network infrastructures, but instead they negotiate resource usage with established network providers.

The advantage to the network operator of these open network interfaces is that the stimulation of service usage results in increased network usage and higher revenues for the transport of the associated network traffic. The network operators are able then to offer a service provision interface to external wholesale customers. This motivation has inspired strong interest in the availability of a standard interface between the computing and telecommunications worlds. This standard interface is an example of an application programming interface (API). For instance, providing customers with direct access to the network to obtain billing information has potential for reducing the large overhead imposed by internal billing systems. This leads to a desire for operating companies to evolve their network structures to be able to interwork with external computer intelligence.

At the same time, the use of object-oriented programming and distributed processing environments (DPEs) has generated great interest, and this has been a prime focus of the TINA (telecommunications information network architecture) project. Since the definitions of the IN SSM (switching state model) in CS-1 and CS-2 the ITU-T IN standards have tended to define 'objects' such as legs and connections, but essentially IN architectures have been described in a functional

---

[16] This aspect of intelligence distribution is known as 'location transparency', because the geographical location of the target customer profile is irrelevant to the requesting service logic. A roaming user's personal profile could be anywhere in the world, so data access procedures must cope with this. The middleware techniques described later in this chapter are designed to provide solutions for this type of scenario.

manner rather than an object-oriented one. Initiatives such as TINA and Parlay are concerned with architectural structures that incorporate the notion of decoupling the network resources that are associated with the transport of information (speech calls or data) from the software that controls the logic of the services themselves.

## 6.3 A model for distributed IN

This section introduces some general concepts for modelling distributed intelligence as a preliminary to describing the relevant industrial standards initiatives, such as the Parlay API.

Figure 6.1 shows a model for the 'decoupling' of the service provision resources from the underlying network resources that control the transport of information (speech calls or multimedia data). The user terminals, which could be fixed or mobile phones or multi-media equipment, are at the bottom of the picture and the services are at the top.

Two different APIs,[17] the network API and the resource API, are illustrated here. The network API is the set of functional interfaces presented to users, third party service providers, applications and their developers and the Resource API is the generic interface to the transport networks.

The essence of distributed IN is the separation of these two domains by the 'middleware', which is a distributed computing environment responsible for providing transport layer resources with access to the required application services. The middleware is a generic distribution facility, providing its users with connections to a resource of the specified type irrespective of the physical details of location or routing. The middleware contains some background components (not shown in Figure 6.1) to carry out the coupling between the transport network user and the required application service. These components include brokering, generic call control and service logic execution environment (SLEE) facilities. Middleware provides the underlying seamless connectivity between the network API and the various resources and transport networks. It ensures that real-time network complexities, and others such as security, load balancing, guaranteed message delivery, etc., are generally taken care of without involving the network API users.

Figure 6.1 shows the middleware interfacing to 'users' (such as application services or transport networks) through dedicated 'plug-INs'. A 'plug-IN' (an intelligent pun) is a modelling representation of an adaptation function that can be thought of as containing the service control, or transport, intelligence particular to

---

[17] API is a term that has transferred to the telecoms world from the computing industry. It describes a concisely defined set of computing methods and rules which allows a computing application to use external resources, or the facilities of other applications. For instance, a mail-handling program on a PC has an API to talk to the PC's modem. It provides the elementary function calls and programming rules. In the broader sense an API specification consists of a set of interface definitions and this, in principle, allows application writers and network operators a free choice of implementation technologies.

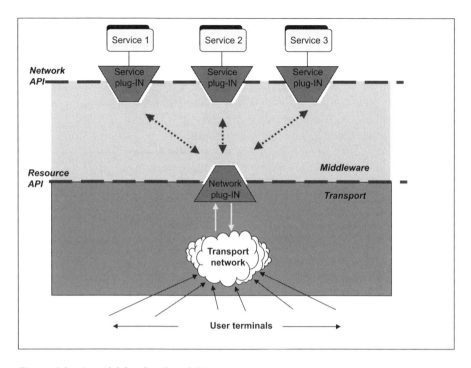

*Figure 6.1 A model for distributed IN*

the target service or transport network. The plug-IN[18] contains the functions that form the bridge between the service-specific, or transport-specific, software and the generic procedures, so the middleware has no notion of the transport technology details. It is equivalent to a 'device driver', where the devices can be signalling interfaces such as H.323, IETF SIP, INAP, CAP, ISUP or ATM Forum UNI 3.1 for accessing the various transport network. For connection to the various application clients and servers across the Network API, plug-INs will ensure the mapping to DPE technologies such as CORBA, DCOM or Java.

Plug-INs can act as clients or servers. Figure 6.2 shows the DPE middleware transporting a request from a client plug-IN to a server plug-IN.

As an example, the network of Figure 6.1 could be a PSTN, in which case the plug-IN that straddles the resource API would be a PSTN-specific plug-IN. This plug-IN then interfaces between the generic transport layer, represented to the middleware as the common resource API, and the PSTN-specific API. The PSTN-specific API could for instance be based on CS-1 INAP. If an IN request (an I_DP) arrives at the PSTN-specific plug-IN then the plug-IN communicates with the middleware brokering function to arrange a connection with an appropriate service

---

[18] The term 'plug-IN' should not be confused with the familiar computing term 'plug-in', which refers to software components used to extend the features of an Internet browser by providing bypass solutions for new browser functions.

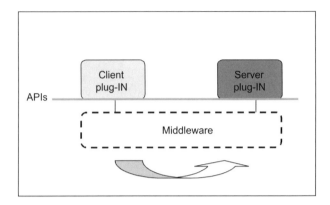

*Figure 6.2   Plug-INs communicating via DPE middleware*

plug-IN. A DPE method call (shown in Figure 6.2) is made to the selected server, which the middleware has chosen to host this service invocation.

The model of Figure 6.1 demonstrates the main characteristics that form the basis of standards such as Parlay for the use of APIs and middleware in telecommunications. This type of model is a consequence of the convergence of the computing and telecommunications industries, where we have seen the emergence of strong drivers to present simple functional interfaces to telecom network application developers and the run-time applications. These interfaces should be independent of hardware platform, operating system or choice of programming language.

Figure 6.3 shows the distributed intelligence model of Figure 6.1 extended to include plug-INs for particular transport network types. Because of the common resource API, which accesses common middleware features and services, the plug-INs for both network types are able to access the same services. Obviously this depends on the network capabilities – a video conferencing service would be available to the broadband multimedia network, but would only be available from the PSTN plug-IN for an ISDN call with suitable terminal equipment connected. Similarly, a voice call diversion service would be available to the PSTN plug-IN as well as to voice calls served by the broadband network.

The broadband network itself could be an IP (Internet protocol) network providing multi-media services including voice, or it might be an ATM (asynchronous transfer mode) network providing ITU-T B-ISDN services.

For the users of the network API, the middleware creates the illusion of the network as a single large, efficient, 'virtual computer', seeming to unify the different transport networks and their corresponding lower level network services.

An important element, already mentioned, is the 'broker', which finds target server objects, of the required type, on request from client objects. Another important entity is a 'naming' service, which assists the broker in finding the correct object. For this purpose, the middleware could typically be based or

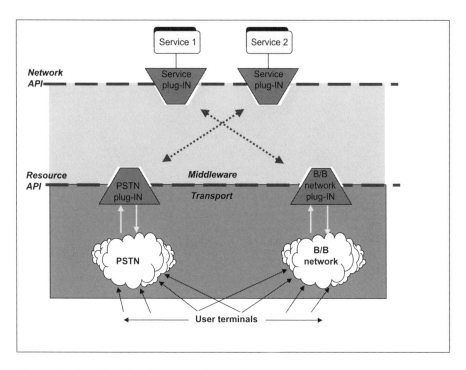

*Figure 6.3 Distributed intelligence and multiple transport types*

modelled on CORBA, although there have been several other candidate technologies for middleware, such as DCOM (Microsoft), RMI (Java) and DTPM (distributed transaction processing middleware).

The broker function is illustrated in Figure 6.4. As can be seen, each of the plug-INs has a 'broker' interface as well as one or more 'transaction' interfaces. The client plug-IN initially uses the broker interface in order to obtain a reference to an appropriate server plug-IN. When the reference has been returned, the plug-INs communicate using the transaction interface.

The notion of the distribution network that has been illustrated in diagrams in this chapter is similar to the model that is commonly used for networking clients and servers in a business environment. The concept of 'network centric computing' allows specialist client terminals to access different services on network servers without needing to be aware of the physical location of the servers. A relevant development in this context is the use of the Java programming language in network computing. Java programs can be quickly downloaded from the servers and run on any client running the Java virtual machine. This type of solution hides the low-level complexities of different computing environments, providing the 'software glue' to allow communication between applications hosted on different equipment.

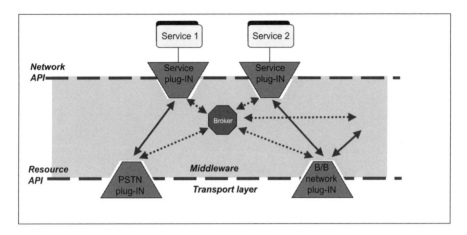

*Figure 6.4  The broker function*

### 6.3.1  Infrastructure APIs

Many APIs are used, at varying detailed levels, in a computing system. Many will be internal and unavailable to third party application developers. An example of a higher level 'infrastructure' API that is likely to be internal to a network operating company's domain is that serving a company's operational support system (OSS), shown in Figure 6.5. Amongst other functions this interface allows call charging to take place.

## 6.4  Object request brokers for network intelligence

The object management architecture (OMA), which was defined by the OMG (the object management group) is based on the object request broker (ORB) component. This allows other elements in the architecture to exchange request and response messages. The principle of the ORB is that it represents ('encapsulates') the entire underlying operating system and network. This is the origin of CORBA (common object request broker architecture).

Most current longer-term strategies for intelligent network architectures incorporate, as a cornerstone, a distributed processing model based on CORBA. The current practical issues with CORBA are scalability and performance. Whilst the OMG has begun work on a real-time version of CORBA, to date a large-scale real-time implementation of CORBA for a telecommunications network has not been built. In fact the first applications of CORBA in telecommunications networks are for off-line, or 'near real-time', management support functions required by the real-time systems.

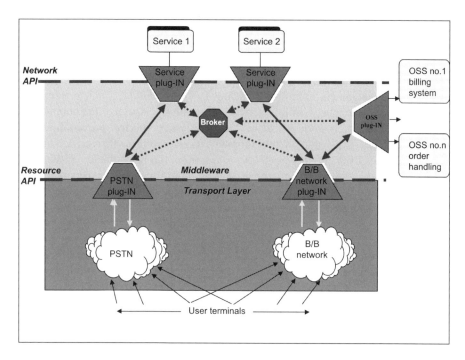

*Figure 6.5 Infrastructure APIs*

## 6.5 The Parlay API

*"**Parlay**, to bet (original stake plus winnings) in a later venture: to succeed in converting an asset into something more valuable."*

Chambers Twentieth Century Dictionary.

*6.5.1 Background*

A leading standard for the network API that has been mentioned more than once so far in this chapter is the Parlay Consortium API. Parlay achieved a high degree of publicity after a revolutionary announcement in 1998 from BT, which was part of the Parlay Group. In this announcement BT indicated its intention to open up its network to third party application developers, using the new Parlay standard. This commitment was made at a time when network growth was seen as critical to the major operators' strategies because competition for network provision was becoming intense.

For a long time there had long been demands from independent enterprise organisations for the ability to offer their own services to their customers in this manner. However, the prospect of laying bare the inner workings of these networks was viewed with trepidation by established operators, who were obviously concerned about integrity breaches. The BT announcement that the network would be opened for outsiders, including commercial rivals, was therefore a bold step, and obviously was not taken before adequate provision had been made for the security and control management issues to be included.

As discussed previously, from a network operator's perspective the commercial driver was that a common standard for third party service provision would stimulate market expansion in telecommunications services, thereby leading to greater revenues from network usage. An easy-to-use, freely available API standard would encourage growth in enterprise start-up companies to the benefit of the telecommunications industry as a whole. Independent specialist service-providers would be able to make their new service offerings available to their own customers using the network operator's underlying voice and data networks and support systems.

The details of the API were published on the Internet (*http:/ /www.parlay.org*), and so were available for anyone to write a service application that would interwork with a public network. The Parlay Group originally comprised British Telecommunications, Microsoft, Nortel Networks, Siemens and Ulticom (formerly known as DGM&S Telecom). These were joined in 1999 by AT&T, Cisco Systems, Ericsson, IBM, Lucent Technologies and Cegetel.

### 6.5.2 What is the Parlay API?

The Parlay API effectively presents the underlying networks as a programmable entity for independent service developers. The particular network technology is hidden from the developer. Parlay was originally based on object technology interfaces from the OMG and so is independent of lower-level messaging technology. For example it is possible to use CORBA, DCOM, JAVA/RMI or any other suitable technology for the distribution middleware.

Parlay comprises a set of APIs that is open, extensible and network-independent. These are intended for use by developers of enterprise-based client applications who wish to write software for services that can use the facilities of participating telecommunications networks. The APIs are object-oriented, and are organised around the functional structure shown in Figure 6.6, which shows the main categories of the Parlay interfaces. Parlay 'service interfaces' provide client applications with access to network capabilities and Parlay 'framework interfaces' provide the backup capabilities such as authentication, load management and fault handling.

The interface classes are specified using UML (unified modelling language), so the Parlay specification is documented in terms of framework interfaces, individual service interfaces, data definitions, class diagrams, sequence diagrams and IDL files. The interfaces are defined in program statements using the OMG IDL (interface definition language). These program statements can be generated automatically in output files from the UML tool.

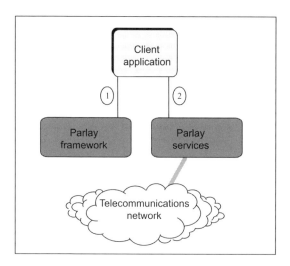

*Figure 6.6   The main Parlay interfaces*

Initially, the prime focus of attention for the Parlay interface definition has been the framework and the services interfaces. These are marked as (1) and (2) in Figure 6.6. In addition to these, work has proceeded on the definition of other public interfaces. These include interfaces introduced to support administrative functions within an enterprise and interfaces (between Parlay framework and Parlay services in the diagram) that allow third party vendors to provide Parlay services. Discussion here is confined to the interfaces shown in Figure 6.6.

◇ The **framework interfaces** include the underlying capabilities needed to provide secure and resilient controlled access to external users. They include the management, authentication and billing capabilities needed as well as the generic features for call and connection services, messaging and user interaction.

◇ The **service interfaces** more specifically detail the functions such as call routing which are needed to build up the services themselves, including access to network capabilities and information.

### 6.5.3  Parlay specifications

The Parlay specification that is in force at the time of writing is Parlay 2.1, although draft specifications for Parlay 3.0 are publicly available.

Parlay 2.1 API specifications are available (from the Parlay Group Web homepage, *http:/ /www.parlay.org*) in the following document groupings:

◇ call processing APIs

◇ connection manager APIs
◇ framework APIs
◇ messaging APIs
◇ mobility APIs
◇ common data & IDL.

The APIs that have a bearing on the IN principles discussed in this book are contained in the 'call processing APIs' document set, which groups the call control service interfaces into the following categories:

◇ generic service
◇ generic call control service
◇ multi-party call control service
◇ multi-media call control service
◇ conference call control service.

Each interface group is described in terms of its interfaces and methods. For example, the Generic Call Control Service [36] has four defined interfaces:

◇ the call control manager interface (has six methods, such as 'createCall()' and 'setCallLoadControl()')
◇ the call control manager application interface (also has six methods, including 'callAborted()' and 'callOverloadEncountered()')
◇ the call interface (has eight methods – examples are 'release()' and 'getMoreDigitsReq()')
◇ the call application interface (has ten methods and is used for call request responses).

The Generic Call Control Service allows applications to carry out call control functions in communications networks, including management of IN services. It is also expected to be adaptable to any required call control technology, including ITU-T H.323, ISUP, Q.931, Q.2931, ATM Forum UNI 3.1 and IETF SIP. The objects employed by the Parlay 2.1 call control interfaces are:

◇ **call** (a relationship between a number of parties)
◇ **call leg** (an association between a call and an address)
◇ **address** (a logical representation of a call party)
◇ **terminal** (an end-point).

These objects will be familiar to the reader from the discussions on ITU-T IN CS-2 in Chapter 5. In fact the call model described for Parlay uses similar notation and manipulation mechanisms as the ITU-T CS-2 call model. The Parlay call party handling procedures are extended to multi-media calls with the Multi Media Call Control Service interface set.

Parlay 2.1 was updated in conjunction with ETSI [37] and JAIN⒯ Community member companies to form the Parlay 3.0 specifications. These are also now part of the Open Service Access (OSA) standards, which was adopted by the 3GPP (3rd Generation Partnership Project) group for UMTS for 3G mobile [38].

## 6.6 TINA

TINA was a relatively long-term, large-scale collaborative project that set out to define a software architecture for future service provision across all types of telecommunications networks. It had its original roots in the US, with INA (information networking architecture), which Bellcore (now Telcordia) instigated in 1990. INA was formulated as a work-program to surpass the IN standards of the day, and it was to incorporate newly emerging distributed processing and object-oriented computing principles.

An important limitation of INA was that it sought to integrate IN with the new regime. Unfortunately this ambitious aim turned out to be impossible to sustain, and INA was overtaken by TINA. TINA, on the other hand, was unhindered by any obligation to be backward compatible with legacy networks, or even intelligent networks. It was driven by the convergence between telecommunications and computing and the aim was to produce an over-arching architecture that would embrace all telecommunications applications.

TINA is rooted in the principles of object-oriented design, autonomous software components and DPE using technologies that had been defined by the OMG for the computing industry. CORBA was a favoured technology for the DPE 'broking' function for TINA. Output from TINA was in turn taken up by the OMG's own telecommunication domain task force.

TINA-C (the TINA Consortium) was formed in 1993 and was disbanded at the end of 2000. It comprised 40 network operators and equipment vendors with a core team working in New Jersey, USA. The aim of the consortium was to produce a document set for the TINA architecture and to work towards easing the path for the uptake of TINA-based products. The TINA-C approach represented a radical review of existing ideas about the evolution of the telecommunications industry. It derived from a general concern that traditional strategic directions were limited and that wider strategic vision would be needed for the new services to run on converged data and voice networks. TINA-C set out to re-use existing standards and other publicly available results, and to produce specifications that could in turn be standardised by the recognised standards bodies.

### 6.6.1 TINA roles and reference points

A logical starting point for understanding the TINA approach is the specification for the TINA business model [39], which is described in terms of actual commercial interactions in a business environment. It describes the separate roles involved. Typical roles described are:

◇ **consumer** – represents the end-user of the service.
◇ **retailer** – concerned with service presentation and interfacing to customers.
◇ **broker** – enables one 'stakeholder'[19] to locate another stakeholder of suitable type. It therefore requires management of an addressing facility.
◇ **third-party service provider** – concerned with production and maintenance of the service elements. If a video shop is an example of the retailer, then the third party service provider (wholesaler) supplies the retailer with his new video-tapes (the service 'content'), and is concerned with issues such as production and supply–maintenance aspects rather than the end-customers.
◇ **connectivity provider** – for setting up and controlling communication paths for content transmission wherever they are needed.

A set of reference points (RPs) is defined in [39], against which TINA conformance can be measured. The roles and reference points are illustrated in Figure 6.7.

As an example, a company in the communications industry will typically be represented by several TINA roles. A video-on-demand retailer, for instance, could also be a third party service provider. TINA therefore defines business administrative domains. The use of the standard RPs within a company's sphere of business is not important, but the standard RPs must be used between communicating domains, as shown in Figure 6.8.

Generally, reference point descriptions are broken down into standard component parts. The 'business part' is to do with the overall scope of the interface, its business purpose and its limitations. The 'information part' describes shared information, such as naming and addressing conventions. The 'computational part' defines the computing object interfaces and the 'engineering part' is concerned with the structure of the DPE nodes, their separation and communication details.

RP specifications operate at several levels, with the interactions broadly grouped into two divisions, as indicated in Figure 6.9. The application-related RP part is concerned with the business, informational and computing viewpoints, whereas the DPE-related part is for the underlying engineering and technology interactions.

*6.6.2 The TINA architecture*

The TINA architecture is made up of three separate sub-architectures. These are:

◇ the service architecture
◇ the network architecture
◇ the computing architecture.

---

[19] A stakeholder is an agent that adopts one or more of the TINA roles.

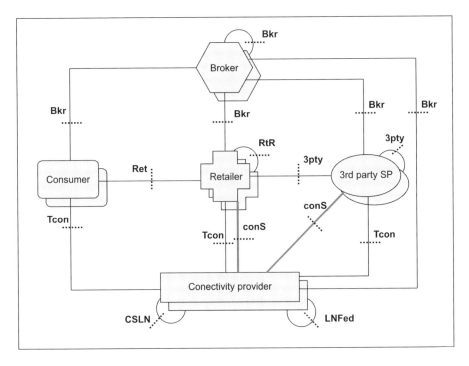

*Figure 6.7 TINA roles and reference points*
*The reference points illustrated in the diagram are:*

| | |
|---|---|
| ***ret*** | *retailer business relationship* |
| ***RtR*** | *retailer to retailer business relationship* |
| ***Bkr*** | *broker business relationship* |
| ***Tcon*** | *terminal connection business relationship* |
| ***ConS*** | *connectivity service business relationship* |
| ***3pty*** | *third party business relationship* |
| ***LNFed*** | *layer network federation business relationship* |
| ***CSLN*** | *client server layer network relationship.* |

Following on from the TINA business model described earlier, an objective of the TINA **service architecture** is to enable the different administrative domains to interact in pursuance of their commercial aims.

The service architecture definition describes how services are designed and used in a TINA system. It defines the software components needed, and explains how the software should be structured in order for it to plug into the overall TINA service management environment. It provides developers with the necessary programming instructions to enable an operational TINA service to be controlled and managed by the TINA environment. The service architecture introduces the concept of

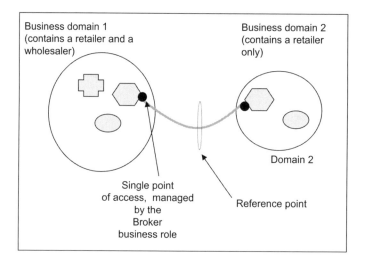

*Figure 6.8   Interaction between business administration domains*

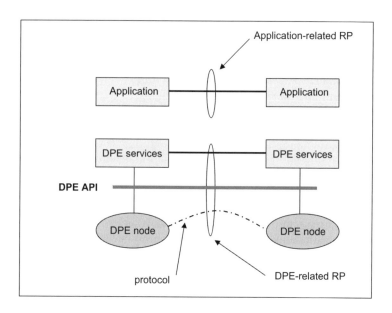

*Figure 6.9   Grouping of RP interaction types*

'sessions', which comprise the related activities in a service invocation. There are four types of sessions:

◇ service session (single activation on a TINA service)
◇ user session (a user's relationship with a service session)
◇ communications session (the assembly of connections required for the service session)
◇ access session (a user's connection to a service session).

As an introduction to TINA service architecture concepts, the 'functional entities' that comprise the TINA model are shown in Figure 6.10 grouped into three domains – those of the 'user', the 'service provider' and the 'network provider'.

Associated with the user domain are 'terminal agents', through which a user accesses services. A terminal agent will exist for every terminal connected to the network, and will be responsible for holding information about the terminal's characteristics such as processing capacity, memory and communications capabilities. This information will be available to the service provider domain.

Associated with the service provider domain are the TINA user agents, service sessions and user sessions. The user agent represents the user, holding information

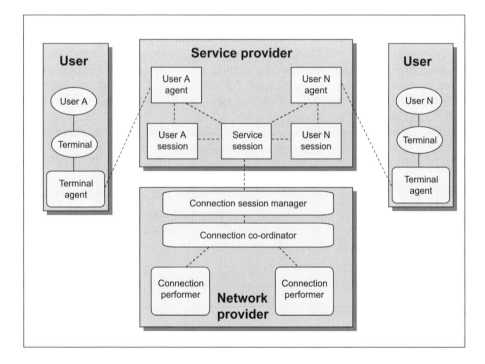

*Figure 6.10   TINA agents and sessions*

such as preferences, subscription records, usage information, password and location, thereby controlling their access to services. There would be a default user agent for unknown users.

When a user invokes a service, a service session is started and adapted according to information held in the terminal and user agents. A user session is also started for each user involved in the session. Amongst the data collected and recorded in a user session is that for subsequent billing. When the user leaves the service session, this billing information would be passed to their user agent for aggregation with other, similar, data.

One of the basic TINA separations in the service architecture is that between 'access' and 'usage', where the former encompasses the interactions for discovering and requesting services and the latter includes the interactions for controlling service behaviour or delivering service content. Access interactions (covering authentication, usage selection, etc.) are needed before the communicating domains can establish the required usage relationships. Usage relationships are for controlling service behaviour as well as maintaining stream connections amongst the service components in order to deliver content information. Service control is to do with establishing the required quality of service, multi-media stream binding, connection control and related topics.

Associated with the network domain are 'communication session managers', 'connection co-ordinators' and 'connection performers'. A communication session manager is invoked by a service session from the service provider domain. With the invocation, the service session also sends terminal capability information, and user requirements for the session. For example, the terminal may be restricted to 64 kbit/s voice transmissions and the user may require a specific quality of service. The communication session maintains the network resources, or modifies them under the direction of the service session (if, for example, an additional user joins the session). The connection co-ordinator arranges for the connection performers to effect the necessary connections.

The **network architecture** is a technology-independent model and it operates at three layers. These are the communication session layer, the connectivity session layer and the network layer.

The **computing architecture** deals with the DPE, which makes the underlying distributed computing equipment behave as if there were a single flexible system across which the applications reside and communicate. The DPE is based on CORBA. The computing sub-architecture also provides the various models, which enable the complexity of the software design to be broken up and viewed from different standpoints.

*6.6.3 TINA specifications*

The final set of TINA specifications and supporting documents is publicly available at the TINA-C Web server (*www.tinac.com*). The document set includes the following specifications, as well as some supporting texts.

***TINA service architecture and specifications***

| | | | |
|---|---|---|---|
| TINA service architecture | Version: 5.0 | TINA-C deliverable | June 1997 |
| TINA retailer reference point | Version: 1.1 | | April 1999 |
| TINA service component specification – Part B (computational model and dynamics) | | | |
| | Version: 1.0 | | Jan. 1998 |

***TINA network resource architecture and specifications***

| | | |
|---|---|---|
| TINA network resource architecture | | |
| | Version: 3.0 | Feb. 1997 |
| TINA network resource information model | | |
| | Version 2.2 draft baseline | Nov. 1997 |
| TINA connectivity service reference point | | |
| | Version 1.0 | Feb. 1997 |
| TINA terminal connection reference point | | |
| | Version: 1.0 | Nov. 1996 |
| TINA network component specifications | | |
| | Version: 5.0 | June 1997 |

***TINA overall model***

| | | |
|---|---|---|
| TINA business model and reference points | | |
| | Version: 4.0 | May 1997 |

## 6.6.4 *Relationship between IN and TINA*

IN has been essentially limited to voice telephony, whereas TINA's ambitions were always wider, being concerned with broadband multi-media, multi-party and information services. TINA is not particularly concerned with transmission details, and is certainly not constrained to services that are restricted to circuit connections. TINA is just as concerned with information and multi-media services using connectionless data streams.

Whereas IN service embellishments start from the 'here and now' of telephone calls, connected via existing telephone exchanges, TINA operates at a higher plane, considering the new services as software-based applications operating on a distributed computing platform. This approach is more concerned with the use of portable code in achieving the required session interactions for the advanced services than with the access practicalities and network technologies, or even the distribution details.

Before the current BICC (bearer-independent call control) standardisation initiative in ITU-T, TINA was recommending a functional split between bearer control and call control, whereas traditional IN places the main functional split between call control and service control (i.e. it preserves call control into the network domain). These differences are illustrated in Figure 6.11.

TINA itself is principally concerned with the service layer, and it defines an object framework for distributed-object implementations of telecommunications

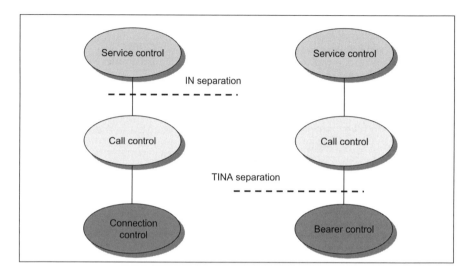

*Figure 6.11   IN and TINA separations*

services of all varieties. Early on it was decided that a sensible migration approach between the different worlds of IN and TINA would be to introduce 'islands' of TINA within existing IN architectures.

*Chapter 7*
# Service examples

*The concept is interesting and well-formed, but in order to earn better than a 'C,' the idea must be feasible.*
—A Yale University management professor in response to Fred Smith's paper proposing a reliable overnight delivery service. (Smith went on to found Federal Express Corporation)

This chapter explores some commonly used examples of the application of IN principles to telecommunications services. As we have indicated previously, IN was primarily intended for voice services, so most of the examples described here are therefore, unsurprisingly, voice features. Many of the examples are variants of the number translation capabilities of IN, but IN number translation has many interesting and revenue-producing guises, as we shall demonstrate in the next few pages.

## 7.1 Simple number translations – Freephone service

The simplest use for an intelligent network is to provide a straightforward number translation service. This really was the way IN started, as we have discussed in earlier chapters. With conventional telephony networking the service intelligence is distributed across the exchanges and the exchange processors pass service information to each other using the common control signalling system. Service and customer data are therefore distributed across the telephony network and conventional (proprietary) exchange management systems are used for data maintenance. The key advantage of using IN technology, however, is that the service logic and customer data is logically held in one place. The management of this data is therefore decoupled from traditional telephony data input systems, and state-of-the-art computing technology can be used for efficient control and fast introduction of new services. Number translation services, where strings of dialled digits are transformed into network routing numbers, are therefore ideal applications for IN.

The success of the Freephone service, which combines these straightforward number translations with the ability for a caller to make a 'free' call (at the recipient's expense), has been world-wide. Freephone dialling codes (e.g. '180x' in the US or '080x' in the UK) bear no relation to the geographical location of the recipient but are convenient to use. In fact, countries that had the foresight to keep letters as well as numbers on their telephone instruments found this factor to be a powerful catalyst in the growth of their Freephone services. This encouraged a source of inventive ingenuity in the development of the services (easy-to-remember 'numbers', such as 180x-AIRWAYS, became powerful advertising aids to market growth).

We looked at the triggering aspect of a Freephone call in Chapter 2. Figure 7.1 summarises the essential network features of the Freephone service.

This is an example of a call to a company which has rented the Freephone number 08081 570980. When callers dial into the company with this number the SSP triggers at the 0808 digits (for example). Then, knowing that the number length for addresses starting '0808' is 11 digits, the SSP waits for a further seven digits before sending a query (a message containing an INAP I_DP operation) to the SCP. The SCP consults its translation database and returns a network number (01632960001 in this case) and instructs the SSP to route the call towards this number in an INAP 'Connect' message. The SSP now recognises '01632' as corresponding to a physical routing, and so is able to complete the set-up towards the call's destination using its normal routing and translation functions, in the same way as it would if the caller had dialled the digits 01632960001 in the first place. However, an important difference is that the call is not charged to the caller, and the resulting call record indicates that a charge needs to be raised against the recipient of the call.

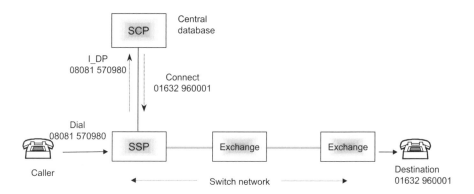

*Figure 7.1   A Freephone call*

In Figure 7.1 we show the scenario where the caller's local exchange has been upgraded to SSP status, but we could equally have shown the second exchange, which is likely to be a trunk level exchange, as the triggering SSP.

We can see from this example that the function of the SCP is to enhance the switch's number translation capabilities. Whilst number-to-address translation is a normal switch function, it would be an expensive and inefficient overhead to load the switch's real-time call-processing software with the functions and data needed for these extra (non-geographic) translations. Conversely, it would be inefficient to distribute the Freephone translation data to hundreds of network switches. This is because these sorts of large-scale network data update operations are slow, labour-intensive and potentially error-prone procedures. Also, larger Freephone customers often have a widely-dispersed and dynamic profile of call reception points, and, through the working day, they may well need immediate changes to their incoming routing translations in order to adjust their response to changing calling patterns. So, having a single point of database update becomes an important advantage for these customers.

### 7.1.1 Inter-network Freephone

The previous example shows the operation of a Freephone service within a single network operator's domain. This next example considers the case of a caller dialling a Freephone number that belongs to another network operator, in the same or a different country. There may, or may not, be co-operation between the intelligent network controls in these two domains.

Figure 7.2 shows the case of 'non-co-operating', independent INs, where SCP processing is invoked in the first network to obtain a translation that can be used to

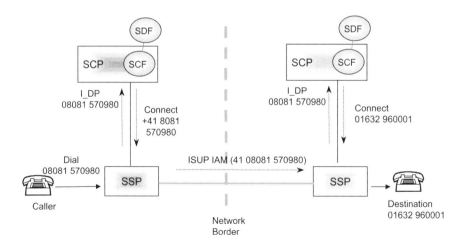

*Figure 7.2   Inter-domain Freephone – non-co-operating INs*

route the call towards the second network. This example shows that the translation process does not alter the Freephone number (which is controlled by the second network operator) but adds a prefix (represented by +41 in this example) that allows the underlying transport network to route the call forward in the right direction. The current example shows ISUP being used as the telephony-level interconnect protocol.

When the call request arrives at the destination network the second network operator's SCP is separately used to find the translation of the Freephone number to the required network termination. The call attempt is then routed towards this required destination.

As an alternative implementation, Figure 7.3 shows the case of 'co-operating' INs, where the first SCP interrogates the second SCP to find the destination routing number directly before sending instructions to the first SSP to route the call.

Overall the processing in this mode of working is likely to be more efficient, as can be seen from a quick glance at the diagrams – a single IN access operation is needed rather than two separate transactions. Furthermore, the 'co-operating networks' approach allows the second operator's SCP to be aware of traffic loading in advance of the incoming calls being offered. An alternative interconnect route, maybe via a third network if necessary, could be specified if the traffic control algorithms in the second SCP are aware of congestion or other service-affecting problems, on the first choice routing to the target destination.

SCF–SCF interface details are included in the ITU-T IN CS-2 recommendations but because there are other commercial issues concerning interconnect agreements, the former scenario (Figure 7.2) is still the more commonly used arrangement in fixed networks. From a standards viewpoint, the diagrams illustrate that the network interconnect could be between SCFs, between SDFs, or between an SCF and a foreign SDF.

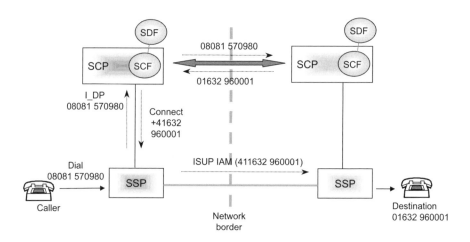

*Figure 7.3   Inter-domain Freephone – co-operating INs*

## 7.1.2 *More complex number translations – Geographical routing*

Once a basic Freephone service is operating, a natural development is to introduce further applications on the number translation theme, making more use of the inherent capabilities of the computer on which the central database is based. Examples are routings based on time of day, day of week or the geographical location of the calling line.

An example of geographical routing would be a national windscreen repair company that offers a single Freephone contact number nationally, with incoming calls always being connected to the geographically closest depot. Another example is the well-known pizza service, where a chain of pizza restaurants advertises a single Freephone number for a home-delivery pizza service and the call is routed to the nearest restaurant to the caller. In the USA this is known as an 'area number calling' (ANC) service.

As before, the SSP recognises the Freephone access code (080X in this case, but could be 1800, or 0500, depending on the local regional numbering plan) and routes a query to the remote database. This time the query includes the CLI (calling line identity). The SCP database recognises, from the Freephone number, that this is to be geographical routing service and so it then obtains the caller's location from the CLI to determine a suitable network destination address for a local agent. The SCP then returns this local agent's network address to the SSP for onward routing of the call.

There can be situations where the triggering SSP does not have a full CLI to send in the IN request message to the SCP. This could arise for instance if the triggering SSP is at the trunk level and is several switches away from the caller's local exchange, which might be an analogue switch that does not support SS7. In this case the normal procedure is for the SS7 network to provide a 'partial CLI', which identifies the digital (SS7-enabled) switch nearest to the calling user. This partial CLI is then delivered to the SCP by the SSP. This will give the service logic some indication of the caller's local area and this can be used instead of the full CLI to find a suitable local agent.

## 7.2 Personal numbering

Personal numbering (PN) is a service that allows users to have overall control of the handling of their incoming calls. The PN subscriber is allocated a 'personal' number, which is likely to be part of a reserved number range so that it can be easily identified by callers as a personal number from the initial digit string.

When callers dial the user's personal number, the service will divert incoming calls to wherever the user wants the calls to terminate. This could for instance be a mobile number, a home number, work number or answering machine. The service maintains a diary with default numbers that can be over-ridden by the user. For instance, a user might want calls directed mostly to his office phone during normal working hours, although whilst travelling to a meeting he might want calls to go to a mobile phone and then to an answering service whilst the meeting is in progress.

These frequent changes to the network destination for calls dialled to the same customer number suggest that an IN architecture with central control of the routing numbers is likely to provide the most suitable implementation. In practice, variants of this service usually have a pre-set background diary profile, with a mechanism provided by the network operator to enable the user to update his profile at short notice.

One automatic and commonly used method is to provide the means for users to enter into dialogues with automatic voice response units prompting the user to enter DTMF digits on the telephone key-pad. In the network this would use intelligent peripheral (IP) equipment and the IN CS-1 'prompt and collect' procedures, described in Chapter 2. In some implementations these updates are more conveniently carried out via an Internet browser and a secure communication with a server that is associated with the controlling SCP.

## 7.3 Incoming call screening

A popular IN application is the screening of incoming calls, where customers need to be able to control which calls are offered to their telephones. Such a service would be attractive for instance for a customer who suffers from a large number of unwanted and unsolicited marketing telesales calls. A 'blacklist' of calling numbers is held in the database associated with the controlling service logic and a terminating trigger (DP 12 on the BSCM in Figure 2.8, p. 29) is set against the subscribing customer's line. Every incoming call to that line causes an I_DP to be sent to the SCF. The I_DP will include the CLI as a parameter, and this will be checked against the barred CLI list in the SCP. If the caller's number appears on the list, a ReleaseCall message is returned to the SSP and the call is cleared. Statistics can be maintained at the SCF, so the subscribing customer can be charged, if necessary, at a rate that is based on the number of rejected calls.

This screening scenario is illustrated in Figure 7.4. The service, or its variants, is sometimes marketed under names such as 'do-not-disturb' or 'choose to refuse'.

Figure 7.4 shows a version of the call screening service where the SCP checks the CLI and instructs the SSP to release the call if the CLI is on the screening list. In practice, the SCP would probably precede the ReleaseCall operation with a PA (Play Announcement) operation so that a suitable rejection tone or automatic announcement could be provided from the SSP's internal SRF before the SSP releases the call.

A more sophisticated variant of the screening service could involve an external SRF provided in an IN IP node. This would allow the SCP to instruct the SSP to extend the caller's connection to the external SRF for more complex screening treatment. For instance, the different levels of screening might apply at different times of the day and customised announcements and mailbox facilities may be required, depending on the screening conditions.

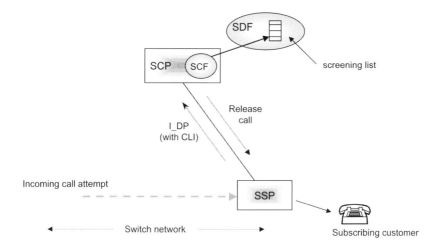

*Figure 7.4   Incoming call screening*

Another variant, shown in Figure 7.5, could be offered where the customer can invoke a 'do-not-disturb' feature, which would result in all incoming calls that are not rejected outright being connected to the IP. The caller can then be invited to enter a 3 or 4 digit 'pass-code' to achieve connection or be connected to the customer's mailbox. If a pass-code is entered by the caller, it could be validated by SRF software, or reported back to the SCF for validation using the Prompt and Collect User Information response message. If the validation is positive, the SCF

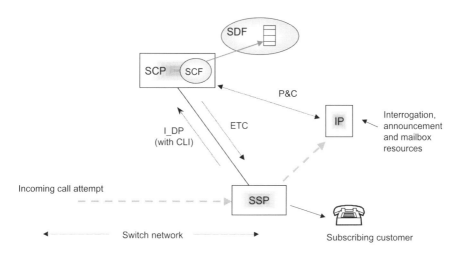

*Figure 7.5   More complex call screening*

would instruct the SSP to break the IP connection and connect the caller to the customer's line.

The message flow for a successful connection after the screening process just described is similar to that of Figure 2.15 (p. 42). The pass-code entry occurs at the 'user interaction' phase, when the caller is in communication with the SRF at the IP.

## 7.4 Least cost routing

In an environment where there is strong competition amongst network carriers for transit traffic, IN techniques can provide an effective solution for a 'least cost' routing service. With this service, trunk calls can be routed over one of several transit networks depending on cost. Figure 7.6 illustrates the operation of the service for a call between customers on two separate local operator networks.

There is competition at the transit level, and so the diagram shows a choice of three different transit carriers. Each is likely to have advantages for different call scenarios, and so the service provided at the SCP needs to be aware of the details of the tariff arrangements for each transit network. There could, for instance, be variations depending on time of day, call destination and call origin.

The service triggers to an SCP at the customer's local exchange, which is the SSP shown in Figure 7.6. This could be a conditional line-based trigger, with a

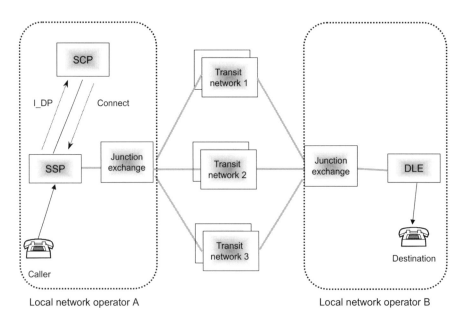

*Figure 7.6   Least cost routing service*

condition that the call triggers if a call is recognised to be a long distance call (e.g. on the condition that there is a leading '0' in the dialled digit string).

The SCP returns a routing prefix, or an outgoing route identifier, which enables the SSP to route the call onward through the required transit network, which relays the call set-up messages to the destination operator's network. The call is then connected to the destination customer in the normal manner.

Providing a fallback option to cover the event that the preferred route is unavailable could enhance this service. An E_DP for a 'route-select fail' (DP-4) event could be set. If this triggers, the call could be connected to an SRF in an IP node and the caller offered a connection over a second choice route. This second option may not be the 'least cost' alternative, but at least the call would be completed – if this is what the customer wants.

This is another example of a service that is dependent on specialist and up-to-date data. In this case tariffs may fluctuate and special price offers may be available at short notice for limited periods. It may therefore be practical for the network operator to out-source the data maintenance function to a specialist service provider. Figure 7.7 shows a scenario where a network operator might offer its own basic service, but may also provide access to a more advanced specialist service operated by an independent service provider. Figure 7.7 shows an internal SDF for the operator's service and an external SCF–SDF interface to the service provider's own SDF, shown in a separate SDP node.

The decision on which data to access could be made on the basis of the service key sent from the SSP in the I_DP message. Alternatively, the SSF could access a customer profile database, which would indicate which service variant has the user's subscription. An advantage of this latter procedure is that the underlying network management need only be concerned with the triggering criteria, and can be kept free from details of the service provisioning.

The SCP application therefore either accesses its own database for least cost routing information or it generates a query to the external database (SDP) which

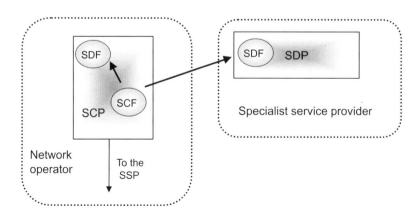

*Figure 7.7   Service data provided externally*

will return the identity of the preferred carrier. The SCP will convert this to routing information and convey it to the originating SSP.

In this scenario the service provider could charge the network operator a per-transaction fee, and the network operator could recover the call costs through normal billing procedures. A different approach would be for the service provider to provide the least cost routing service directly to the customer, making its own provision for billing, using the CLI or PIN verification for identification on a call-by-call basis. In this case the network operator's SCP would just operate as a gateway, providing access, control of SRF resources and an onward routing service.

## 7.5  A virtual private network (VPN) service example

To further illustrate triggering principles we now examine a simple VPN service, which is illustrated in Figure 7.8. The features offered by a VPN service typically include:

- [ ] abbreviated dialling
- [ ] reduced call charges for 'on-net' calls
- [ ] a call forwarding service
- [ ] a mailbox service.

In this example the triggering criteria is arranged such that when a user dials the digit '7' the SSP triggers but waits for three more digits before sending the request to the SCP. '7' is the prefix for the VPN numbering scheme and users have three

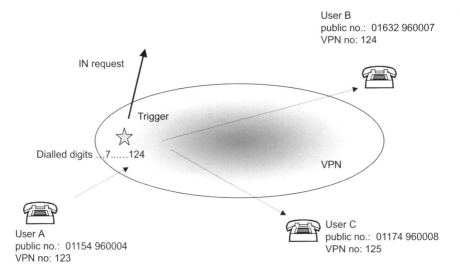

*Figure 7.8   A VPN service*

digit private addresses. These are in addition to their public addresses, as shown in Figure 7.8. This trigger is equivalent to the 'customised dialling plan trigger' in the US AIN specifications. The service operates as follows:

User A dials user B's VPN number, which is '124', preceded by a '7', which is the pre-assigned identification digit for VPN users calling each other. The SSF assembles an I_DP message, ensuring that it contains the digits that user A has just dialled, as well as A's own public number (01154 960004). The I_DP also contains the service key (SK) that is associated with the particular set of triggering conditions provided in the SSP for this VPN service.

The I_DP is despatched to the SCP. The ensuing transaction is shown in Figure 7.9. The request is initially handled by the portal functions of the SLEE (service logic execution environment, which was described in Chapter 2) at the SCP. The SCP as illustrated has four different service logic programs (SLPs) running, which are accessible through the SLEE. The SLEE uses the SK parameter of the I_DP to identify the appropriate SLP for handling the request.

The network operator will be providing VPN services for many customers, and each one is a separate and independent virtual network. The particular VPN associated with a new call is identified by the network CLI (i.e. the advertised public number in this case) of the user who has called. The VPN access digit ('7') indicates that the following three digits ('124') will be an address for another user

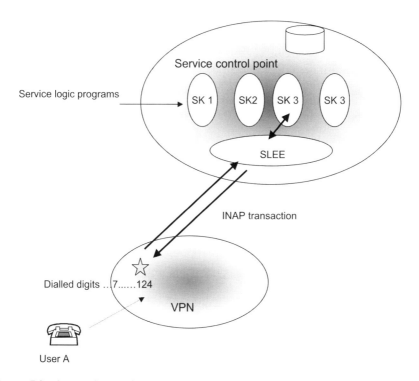

*Figure 7.9   Service logic selection*

on the VPN. Having established the context, the SLP will consult a translation database in order to find a translated number for the onward routing. The database consulted can be local to the SCP or remote, or even under the customer's control.

When the service logic has located the database entry and has found a corresponding network destination number ('01632 960007' in this case) the SCP returns an INAP 'connect' operation to the SSP. The Connect operation normally contains the required public network address ('01632 960007') that is used by the SSP to continue with the call set-up.

However, if the called user has a call-forwarding (unconditional) feature active, then the IN service logic would have returned the network number for the call-forwarded destination. So, if user B is currently forwarding incoming calls to user C, the address string returned in the Connect message would have been '01174 960008'. If the *conditional* call-forwarding service is active, such that calls are forwarded if the user is not available ('no reply' or 'busy'), the IN service logic will precede the earlier Connect operation with E_DP-arming operations. The SCF can arm for the network events 'busy', 'no reply' or 'route unavailable', using 'request report BCSM' operations with 'event' parameters set to 4, 5 or 6. The message sequence diagram of Figure 7.10 illustrates the action taken when one of these E_DPs is encountered.

In order to specify a special charging rate for the VPN service, the SCF also sends an FCI ('furnish charge information') operation. This allows service-related information to be temporarily stored in a working record in SSP so that, at the end of the call, an appropriate charging event record is output to the off-line billing system for consequent pricing and billing.

Figure 7.10 illustrates a typical SSF–SCF message interchange for the example IN VPN service activation. The three messages from the SCF (request report BCSM, FCI and Connect) could be sent as separate SS7 messages, as shown in

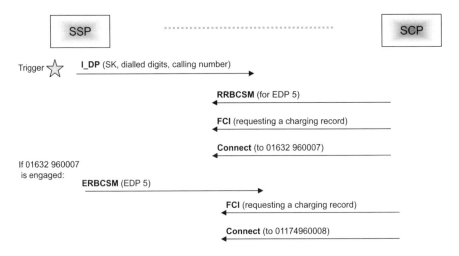

*Figure 7.10   Message flow for a VPN call*

Figure 7.10, but alternatively a single message could be used for all three, leading to greater messaging efficiency.

If the called customer answers, there is no further SCP involvement, and the INAP transaction is closed. However, if the destination is busy, as is the case in the example of Figure 7.10, the SCP will receive an event report BCSM operation, in response to the request report BCSM, and further SCP processing is needed. The SCP will then instruct the SSP to route the call to an alternative destination, as shown in the message sequence.

This service example illustrates some of the principles. In most cases there will be a richer service specification, requiring a more complex interchange of messages. For example, it is likely that E_DPs for other call-failure conditions (e.g. 'no reply', or faults or congestion) will be armed, requiring appropriate SCF logic to be executed when these are triggered. Also a mailbox destination would normally be included as an alternative destination for calls that fail to reach their intended destinations.

## 7.6 A directory enquiry call completion service

This is an example of an IN-enabled directory enquiry service that offers automatic connection to the required destination number when the enquiry operator has found the required number.

The example illustrates ITU-T CS-1 procedures and messages for the SCP-controlled automatic connection of a call to IN IP devices. This is in order for the SCF service logic to collect further information from a caller in order to progress a call. Typically the extra information is collected by the IP, using voice guidance. The caller is invited to enter DTMF tones from key-pad sequences, or to respond verbally to the automated voice unit.

As with the other service examples, the procedure illustrated here is just one of a number of possible ways of arranging the service logic support for a network service. For example, for clarity of explanation a single SCP is shown as the host for the required service logic for this service. In practice, the constituents of this SCP may well be provided in a geographically distributed computing system, but the network control principles illustrated here remain fundamentally similar.

### 7.6.1 IN distribution of operator calls

In preparation for describing the ensuing stages in the call completion scenario, Figure 7.11 shows the basic arrangement for accessing an enquiry operator from a common access number. Typically a short well-known digit string is used to access services such as directory enquires – this illustration uses the digits '123'. Enquiry operators will typically be geographically distributed over a wide area and the controlling service logic needs to arrange for a new call to be routed to a suitable free assistance operator.

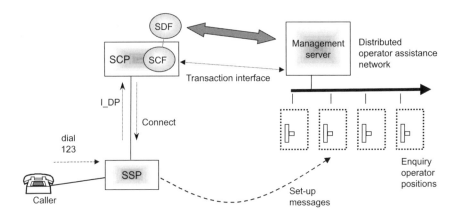

*Figure 7.11   A directory enquiry call*

Commonly, dedicated distribution and local queuing systems are employed for operator assistance centres. If an IN solution is used, these systems must either be provided within the IN service logic in their entirety, or else there must be interworking between the overall IN logic and the local management servers. This is to ensure that the IN systems have up-to-date information available on termination availability, and to update the SCF with results of individual searches. The latter is the scenario chosen for the current example, and the scenario is illustrated in Figure 7.11. These local management servers control local call centres (and possibly home-working groups).

As well as performing normal housekeeping and supervisory functions, the management servers also continually update the service data function (SDF) at the controlling SCP with information on availability and preferred choice of selection of the network numbers corresponding to the enquiry operators. This provides the SCF service logic with continual real-time information to enable it to provide fast responses to incoming requests.

Triggers are provided at SSPs, which can be at the local or trunk level in the network, so when a customer dials '123' to access the service, an IN request message (I_DP) is sent to the SCF to obtain a routing. The SCF's choice of network routing will depend on the CLI (sent from the SSP in the I_DP) and on information available in the associated SDF. In reply to an I_DP it responds with a 'connect' message instructing the SSP to route the call to a suitable enquiry operator's network number, in the customer's local area if necessary.

### 7.6.2 Automated response and call completion service

As human labour costs rise relative to costs of equipment and software, the business imperative with enquiry operator procedures is the trimming down of the overall call-handling time in the enquiry operator procedures. One or two seconds' saving here and there on enquiry operator services using automation can make a big financial difference to a network operating company's balance sheet. One way to do this is for the enquiry operator to pre-record a salutation (such as 'Good morning. Which name and area do you require?'). This is then played to the caller by a voice response unit (VRU) at the start of every call, so the enquiry operator is then able to join in the call when the caller is giving the details.

Another commonly used way of saving time is, after the trace has been completed, for the required number to be given by an automatic announcement, freeing the operator to move on to another enquiry after finding the number. The call completion service can be offered at this point by another VRU, and if the caller agrees, the call can be established automatically by the SCF.

The illustrative sequence here (Figure 7.12) is for a call that passes through each of these stages. The overall process therefore requires a series of connections to a VRU, then to an enquiry operator, then another VRU and finally to the destination number. The VRUs are numbered 1 and 2 in Figure 7.13, but obviously they could be the same unit. They are shown separately here to illustrate the principle of serial interconnections that is characteristic of IN CS-1 INAP procedures – each connection is cleared as the next is established.

Connections are numbered in Figure 7.12 as follows. Connection 1 plays the initial salutation and leads into the main interaction between the caller and the enquiry operator in connection 2. Connection 3 to the second VRU is to tell the

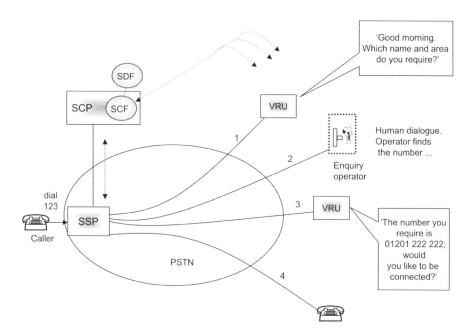

*Figure 7.12   A 4-stage directory enquiry service*

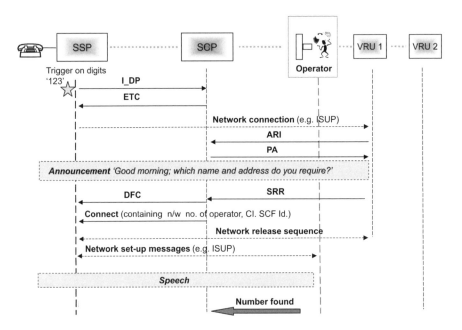

*Figure 7.13   Call completion call flow, part 1*

caller the number that has been found and to offer the call completion service, and connection 4 is the actual call to the destination number.

A message flow that will achieve the network control for the call as far as the enquiry operator interaction (connection 2) is shown in Figure 7.13.

When the caller dials the access code for directory assistance triggering occurs as previously described, but this time the SCF instructs the SSP to route the call temporarily to a VRU using the ETC procedures for IN IPs which were described in Chapter 2.

The call then arrives at the VRU and is connected using normal ISUP procedures. Meanwhile the VRU's SRF uses the SCF Id. parameter to find the signalling location of the controlling SCP. The SCP had sent the SCF Id. parameter, with the correlation Id. (CI on the diagram), to the SSP in the ETC operation, and these parameters were then relayed to the VRU in the network signalling address message.

The VRU is therefore able to send a query message (ARI operation) to the SCF, including the correlation Id. parameter. The SCF uses the CI parameter to bind this request to the earlier transaction that it had opened with the originating SSP. The SCF returns the 'Play Announcement' instruction to the SRF. This includes details of the announcement that should be played to the caller. The VRU equipment duly plays the requested announcement and reports completion of this task back to the SCF using the SRR ('specialised resource report' operation).

The SCF now instructs the SSP to clear down the speech path to the first VRU and to set up a new connection to a free enquiry operator. This uses the DFC and connect operations. This call scenario assumes the existence of a background relationship between the controlling IN SCF and the management servers for the enquiry operator call centres, as shown in Figure 7.11 above. This keeps the SCF up to date with information on the network addresses of available enquiry operators. The SSP now releases the speech path to VRU 1 and sets up a new path to the nominated enquiry operator just in time to hear the caller respond to VRU 1's earlier salutation, enabling the enquiry operator to start the number trace.

Having located the required number the operator presses a keyboard button which launches an information message, containing the required number, from the management server to the SCF. This uses the (non-IN) transaction interface in Figure 7.11 to send a 'number found' message to the SCP. The operator will then be free to drop out of this enquiry to move on to other work. The operator's server locates the correct target SCF, using the SCF identifier and CI, which had been sent forward in the network set-up message (e.g. ISUP IAM), which had been received in turn from the SCP in the connect message.

The message sequence continues in Figure 7.14. On receipt of the 'number found' message, the SCF releases the operator connection, using the DFC (disconnect forward connection) message and instructs the SSP to establish a new connection to the second VRU. Again, the SCF uses the ETC operation for this,

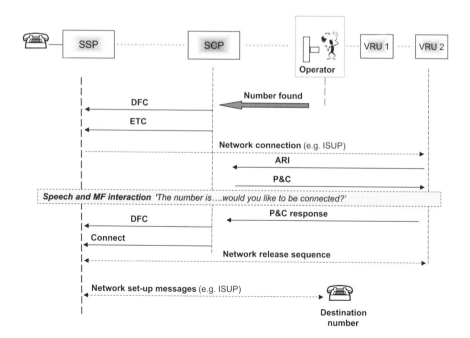

*Figure 7.14   Call completion call flow, part 2*

sending the SCF ID and CI parameters for the same transaction binding and SCP identity reasons as before.

VRU 2's role here is to announce to the caller the number that had been found in the preceding interaction by the enquiry operator and to offer the call completion service. This invites the caller to respond verbally, or to enter DTMF digits to indicate acceptance or otherwise. Because more is required of the SRF in VRU 2 than was required of VRU 1, the SCF this time replies to the VRU's ARI with a 'prompt & collect' (P&C) instruction rather than the passive PA request as before. This time the SCF asks the SRF to collect a response from the caller, either via DTMF digits or voice recognition equipment.

The caller's response is returned in the P&C Response operation. If the caller has accepted the offer of automatic call completion, the SCF once more instructs the SSF to arrange for the disconnection of the temporary connection to the VRU and sends a Connect operation for the SSP to make the final network connection to the required destination number.

From this point on, the call is routed and connected to the required destination number in the normal way. The SSF–SCF transaction could be terminated at this point, or it could be kept open and a mid-call trigger armed, and the call could be re-presented to the operator if necessary.

In practice there are most likely to be other INAP messages used; we have just shown the ones needed to demonstrate a skeletal service. For instance, charging has not been considered; an FCI could be sent with each Connect and ETC message, this would cause the SSP to generate a billing record, if required, for each of the three separate connections used by the caller.

## 7.6.3 Call completion service summary

This example has aimed to illustrate the 'behind-the-scenes' workings of the intelligent network using various CS-1 INAP operations to effect the series of connections needed to make up a complex service of this nature. The actual connection methods used in physical implementations vary widely – we have chosen to illustrate the use of standard Core INAP operations for all interfaces except the SCF to operator management service. However it is common for proprietary optimisations to be included to avoid messaging delays.

Obviously there are stringent messaging performance targets to meet because the various connection changeovers need to appear instantaneous to the user, and delays of more than a few hundred milliseconds at any stage will become noticeable. To achieve the serial coupling of the different service logic functions throughout, the 'correlation Id.' (CI) parameter is essential, as is the SCF Id. parameter, which serves to complete the triangular relationships between SCF, SSF and SRF. This CI parameter appears as an optional parameter in several INAP operations, particularly the Connect, ETC and ARI, which are used here. However, some telephony signalling systems in force today (e.g. TUP derivations and early ISUP versions) do not have standard mechanisms implemented for conveying CI and SCF ID between the SSP and the IP, and so ingenuity is often needed in specific implementations to find ways of re-using existing signalling fields.

Because the majority of today's IN implementations are still based on the CS-1 core INAP procedures, we have illustrated these CS-1 procedures for this service. In fact, this type of service is likely to be more suited to the CS-2 leg-control procedures, because of the numbers of connections and legs involved and the time constraints. Call party handling (Chapter 5) techniques have a big advantage with services where these sorts of time-critical call-leg manipulations are not necessarily serial, and so are handled more efficiently.

## 7.7  Call gapping

Independently of IN, call gapping is a commonly used flow control technique in telephony networks. It is therefore naturally adopted as a mechanism for load control protection at IN service control points and other vulnerable IN nodes that might be at particular risk where they are centrally provided resources.

CS-1 specifies a flow control procedure where an SCP is able to send call-gap request messages to an SSP in the event of resource overload risk in the components of the SCP. The core INAP operation that is sent from the SCF to the SSF in order to reduce the rate of service requests is the 'CallGap' operation. CallGap will instruct the SSF to gap on destination number or service key, or both, and to gap calls for a specified duration with a specified minimum inter-call interval. The INAP CallGap operation is sent to an SSP in a response message for a currently open transaction.

An example is shown in Figure 7.15. In this illustration a call triggers and the SCF returns a translated number for routing in a 'connect' operation, and an FCI operation to convey IN charging information via the SSP to the off-line billing centre. However, the SCF has detected a high level of calls to this particular service, which is identified by the service key value, and so it returns a call gap instruction for this service key value. The gapping does not apply to the current request – it comes into force after this request has been dealt with.

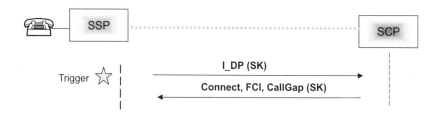

*Figure 7.15   Use of the INAP CallGap operation*

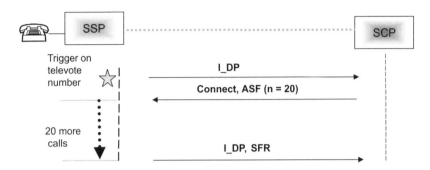

*Figure 7.16   Activate service filtering example*

## 7.8  Activate service filtering

The ActivateServiceFiltering (ASF) operation was designed to ease the load on the SCP in the event of high-volume televoting services. This procedure devolves a large part of the mechanics of the service handling to the SSP by instructing it to only trigger to the SCP for this service every $n$th call. The value of $n$ could be any value within the implementation limits, but might commonly be between 1 and 1000 depending on the dynamics of the service.

Details of the votes are maintained in the SSP in counters, which are defined in the ASF operation. The ASF also specifies the action to be taken by the SSP in terminating the televote calls, in terms of playing announcements or tones to the callers.

Figure 7.16 shows an example where $n$ is set to 20. The results are reported after the next $n$ calls in a ServiceFilteringReport (SFR) operation, which can accompany another I_DP operation.

The issue of charging for the service must of course be settled. The normal procedures allow the SCP to return an FCI operation (e.g. with a connect operation) in response to an I_DP. However, in the case where the SCP has instructed the SCP to take autonomous action, such as connecting the caller to a standard announcement, there may be a need to generate a billing record to the SCP's specification for each of these individual calls. Alternatively, the billing could be left to normal procedures, where typically a call record will be generated by a local exchange for every (apparently) successful call.

## 7.9  Simple CTI (computer–telephony integration)

CTI is what it stands for – the integration of computer technology with telephony. A typical application is that employed in a call centre, where the CLI is used to present customer information on an agent's computer screens whilst simultaneously presenting the associated call through the same agent's headset.

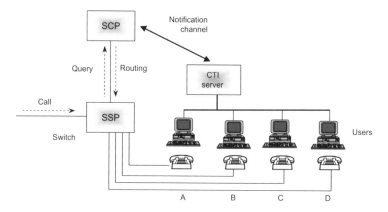

*Figure 7.17   CTI call pick-up*

Figure 7.17 shows an example of a simple CTI-group function, where waiting calls can be directed to alternative agents. The terminals represent four users of a CTI call pick-up group, each having a telephone and a computer terminal. A call is routed through the network destined for telephone D and a terminating trigger ('DP 12 – T_Attempt_Authorised') is encountered.

The SCF processes the I_DP, which includes the dialled number, the CLI, the DP number and the service key value, which enables the SCP to identify the SCF application logic that handles the call pick-up service. The SCF service logic sends a message on the notification channel to the CTI server containing the CLI. The CTI server is able to use the CLI to access records about the caller, and these can be displayed on user D's computer screen while the caller is waiting for an answer. The SCF returns an instruction (containing connect and requestReportBCSM operations) to the SSP to attempt a call connection to D and to arm event triggers (E_DPs 13 & 14) to cover 'no reply' or 'busy' conditions.

In this example the call is offered to D but is not answered, so after the pre-set time period the SSF activates the 'no answer' event trigger and notifies the SCF, using the Event Report BCSM operation. The SCF then accesses information in a local or remote SDF (which could be the CTI server) about other destinations that could accept the call. Having found an alternative destination number, say B on the diagram, the SCF instructs the CTI server to transfer the CLI information from D's computer to B's computer. At the same time the SCF instructs the SSP to clear the call attempt to D and to make another call attempt to B.

The communication between the SCP and the CTI server (the 'notification channel') could be a proprietary interface, or it could be based on IN CS-1 and CS-2 SCF-SCF or SCF-SDF protocols. A key benefit of using IN-based solutions here is not so much the call deflection capability, which many PBXs possess anyway, but that the group members could be geographically remote from each other spread across a wide-area CTI network.

## 7.10  CAMEL calls

### 7.10.1  An outgoing CAMEL call

Figure 7.18 illustrates the call set-up messages for an outgoing mobile call, using a CAMEL IN request, to a PSTN user. This example call flow is for a number translation service. It is concerned only with the call control signals; it is not concerned with the sequence of preliminary signalling needed for the mobile station to obtain a channel from the BSS, or with the authentication dialogue that takes place with the HLR and VLR.

When a mobile station originates a call, and a set-up message arrives at the MSC from the BSC (not shown in the diagram), the MSC interrogates its associated VLR, using a MAP 'send info for outgoing call' message. This message contains the following key parameters:

• the requested call destination, in the form of an E.164 address
• the requested service, which in this example is just voice.

On receipt of the 'send info for outgoing call' message the VLR will check the caller's profile information, which would already be lodged in this VLR as a result of a previous location update. In particular it checks that this caller is authorised to make calls and that call barring does not apply to the requested destination address.

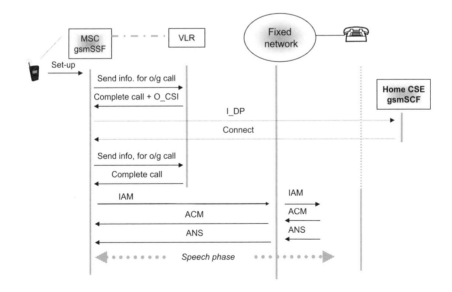

*Figure 7.18   An outgoing CAMEL call*

When the VLR checks the profile it will also discover that the CAMEL subscription information (CSI) mark is present, because the caller is a CAMEL customer. The CSI in the subscriber's data profile in the VLR will have been updated from the HLR. The VLR then returns a 'complete call' message that includes an O_CSI (originating CAMEL subscription information) indicator. O_CSI will identify the particular CAMEL service environment (CSE) for this subscriber.

The home CSE could now be in a foreign mobile network – it makes no difference because CAP (CAMEL application protocol) is an international standard and the gsmSSF does not have a particular relationship with any gsmSCF.

As well as the address of the target CSE, the CSI also contains a service key so the destination gsmSCF knows which services are to be triggered. This is an interesting extension of the CS-1 service key (Chapter 2), which is chosen as a result of the triggering criteria in the SSF. The gsmSSF, however, has no knowledge of the trigger 'conditions', so it does not know which services are to be triggered in individual service invocations.

If the CSI had not been present in the caller's profile information, the VLR would have checked whether conditional barring was set. Because this is a CAMEL customer, the ultimate destination may be changed as a result of the consequent CAMEL service request. The VLR is now only concerned at this stage if barring of **all** calls applies – in which case the call is immediately rejected.

When the gsmSSF receives the 'complete call' message with the CSI, it assembles an SS7 CAP message (to be carried by TCAP and SCCP) containing the CAP I_DP operation. The address (of the home CSE) for the message is obtained from the CSI. The CAP I_DP parameter set would typically include:

◇ service key
◇ calling party number
◇ calling party category
◇ bearer capability
◇ called party number
◇ BCSM event (DP number)
◇ IP SSP capability
◇ caller's IMSI (international mobile subscriber identity)
◇ call reference number
◇ MSC address.[20]

When the gsmSCF at the user's home CSE has processed the I_DP contents, it returns a CAP 'connect' message to the MSC (containing the gsmSSF). This contains a new call destination address to be used by the MSC for this call.

---

[20] It can be seen that, apart from these three last GSM-specific parameters, there are strong similarities between the CAP I_DP and the INAP I_DP. More information on CAMEL standards is given in Chapter 3, and an overview comparison between mobile and PSTN signalling is in Chapter 3.

At this stage a second check with the VLR is carried out, using the MAP 'send info for outgoing call' to ensure that this user is not barred from calling the new destination number. If this check passes, a second 'complete call' instruction (without CSI) returns to the MSC, which sends an ISUP IAM across the network towards the required destination. Figure 7.18 assumes that the next MSC is in fact a gateway MSC, and that the call is destined for a PSTN user. The call set-up and clear-down sequences then proceed in the same way as for a fixed network SS7 call.

## 7.10.2 An incoming CAMEL call

Figure 7.19 shows an incoming call from a fixed network user to a CAMEL mobile subscriber. In this example a CAMEL IN request is made before connection. This allows the CSE to run CAMEL-based call screening services (for example) before authorising the gsmSSF to connect the call. This illustration shows the SCE instructing the gsmSSF to connect the call to the subscriber.

The call enters the mobile network via a Gateway MSC, which queries the HLR in order to find out which MSC to direct the call towards. If there is more than one HLR, the G-MSC, which (in this example) is also the gsmSSF, will identify the

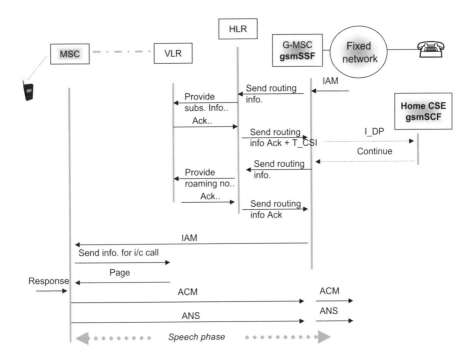

*Figure 7.19   An incoming CAMEL call*

correct one from the called address. This query uses MAP, and, as shown in Figure 7.19, the message used is a 'send routing info.' message.

The HLR will first check that calls for this user are not being forwarded, and will then send a MAP 'provide subscriber info.' request to the last-used MSC/VLR for this mobile subscriber. For non-CAMEL subscribers, the HLR would use the MAP 'provide roaming number' request. However, a different request message is now appropriate because the gsmSSF needs the CAMEL subscriber's location and status details, which need to be sent in the CAP I_DP message to the home SCE at the gsmSCF.

The VLR returns the required details and allocates a temporary mobile station roaming number (MSRN). This information is returned to the G-MSC via the HLR in the acknowledgment messages. The MSRN is subsequently used as part of the destination address in routing the call to the MSC currently serving the mobile. The MSC associates the temporary MSRN with the mobile station, after which it frees up the MSRN for re-use.

The G-MSC (the gsmSSF) next sends an SS7 CAP I_DP request to the CAMEL subscriber's home CSE. The destination address (for the home CSE) for the message is obtained from the information returned in the previous MAP response from the VLR.

After the gsmSCF has processed the I_DP contents and checked the screening conditions it returns a CAP 'continue' message to the gsmSSF, instructing it to carry on processing the call from the point it reached when it sent the I_DP to the gsmSCF. The call then proceeds as a normal incoming GSM call. The G-MSC sends an ISUP IAM, using the previously allocated MSRN, and the target MSC sends a paging signal for the mobile in its last-known location area, then waits for the response from the mobile station. When this is received the call can be connected.

### 7.10.3 Pre-pay mobile service

The 'Pre-pay' mobile service is an example of how IN features can be translated directly into results in the marketplace. Pre-pay was introduced originally as the poor man's mobile phone service – a second class service for users who didn't have the financial basis, or credit worthiness, to enter into a fixed minimum period contract and receive monthly bills. In fact Pre-pay has now turned into a mainstream service with big advantages for users and operators. The overheads of contracts, credit checks, bills, minimum contract periods and monthly network access charges become unnecessary. Bad debt is not possible because airtime is paid for, simply and electronically, in advance. As well as reducing the mobile operators' management overheads, these features expand the market footprint because many potential customers had been deterred by the prospect of contracts and direct debit arrangements. As a result, mobile phones have become affordable commodity items, available from supermarket and other popular outlets.

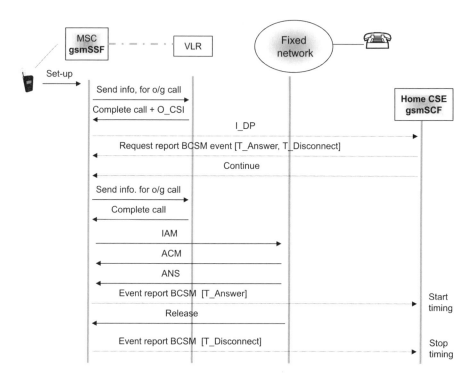

*Figure 7.20   An outgoing pre-pay CAMEL call*

The CAMEL system has provided the Pre-pay service with an inspired technical solution. With CAMEL, all Pre-pay calls involve a request to the CSE, which controls the credit checking and call monitoring. The CSE is able to instruct the gsmSSF if the caller is running out of credit so that the caller can be informed using an automatic announcement. Figure 7.20 shows the messaging involved in a typical Pre-pay call.

As in Figure 7.18, the MSC–VLR interchange results in an I_DP being sent to the home CSE, which recognises the user as a Pre-pay subscriber and returns a request report BCSM event message, arming the gsmSSF for the two events T_Answer and T_Disconnect. This is followed by a continue message, assuming there are no number translation or call barring services applicable, and that the caller has sufficient pre-paid credit in his account. When an ANS (answer) message is returned from the far end network, the CSE is informed, via an event report BCSM message, and it can start timing for the call. At the end of the call another event report BCSM message is sent to the CSE, which can stop the call duration timer and reduce the user's available credit appropriately.

Had the user run out of credit during the call, the gsmSCF could have sent a 'release call' instruction to the gsmSSF and the MSC would have terminated the call.

## 7.11  IN control for Internet dial-up access

An important application for IN today is in the management of the routing of dial-up access to Internet service providers (ISPs). Where the basic access to the IP network is provided by dial-up connections, IN control can substantially streamline the delivery of dial-up data calls to appropriate modem banks in one of the ISP's points of presence (PoPs).

An example of this is shown in Figure 7.21, where an ISP has two PoPs containing modem banks that can handle incoming dial-up calls for Internet access. The caller dials an Internet access number, which would typically be via a Freephone or premium rate number controlled by an IN platform. The IN SCP therefore needs to provide a translation to a network address that will deliver the call to a suitable modem bank.

In the example shown, the call is offered to PoP1, but all ports on PoP1 are busy and so it is unable to handle the call. The IN platform is able to intercept this call failure, by arming E_DPs (in the way we have demonstrated in previous examples in this chapter) and re-routing the call to a secondary PoP (PoP2) which has spare ports. In this way, a call that would have otherwise failed has been connected to the IP network.

As the IN platform is routing the call to the PoP, it is able to apply conventional voice network control features such as proportional routing, call gapping, etc. An obvious enhancement would be to provide a data link between the SCP and the ISP's own management system so that SCP routing decisions can be moderated by information or instructions from ISP-controlled service logic. An example of a standard ITU-T IN component that could be used for this type of application would be the CS-2 intelligent access function (IAF), which is designed to provide for linkage between non-IN and IN network structures.

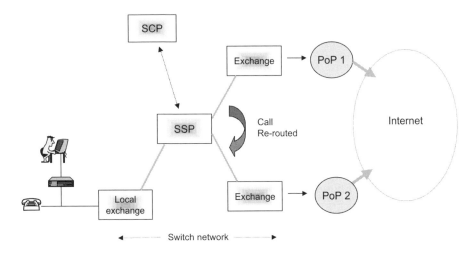

*Figure 7.21   IN control of Internet dial-up access*

## Chapter 8
# Concluding summary

*Everything that can be invented has been invented.*
—Charles H. Duell, Commissioner, U.S. Office of Patents, 1899.

### 8.1 IN today – How did we get here?

Today, intelligent network principles are well established in most large fixed public networks, and are controlling many popular high-volume services providing backbone revenue streams for the large network operators.

Most intelligent networks in operation today have their basis in the ITU-T's highly successful CS-1 standard, or the (broadly equivalent) AIN standard in the USA. In order to illustrate the potential for the evolution of the current implementations, we have, in this book, mapped out the range of intelligent network recommendations that have been produced by the ITU-T, which is the main international standards body for telecommunications. Regional variants of IN, such as AIN in the USA and ETSI's Core INAP in Europe have been based on the ITU-T's recommendations.

ITU-T CS-1 was a major early success for IN, but the next standards landmark, CS-2, represented a considerable expansion of detail that has so far proved to be too much for many network operators, although this applies less to newer operating companies who do not have such large legacy overheads. CS-3 consolidated on some of the CS-2 recommendations and started to venture towards broadband networks. We have seen that the recently released CS-4 documents reassess the role of IN against a background of converging PSTN and IP networks. CS-4 seeks to reconcile traditional IN with the IETF initiated activities such as PINT and SPIRIT for supporting services such as click-to-dial and Internet call waiting. CS-4 also tracks the API Open Service Access initiative for interfacing IN with external service provider applications.

However, this book has been primarily concerned with implementations based on the ETSI Core INAP standard because this has become a widely popular variant, and was in fact fed back into the ITU-T to be adopted as a refinement (CS-1R) to the early CS-1 ITU-T IN capability set. For this reason this book has concentrated primarily on the practicalities of the 'here and now' of IN as it is currently operating, rather than indulge in speculation of how evolving IN principles might be implemented in future mainstream networks.

We have also provided some coverage on the origins of IN, from the early pioneering Bellcore work in the 1980s (and before). This is because it is important to understand the context into which IN grew in order to understand the possibilities, as well as the limitations, of the technology.

During the 1990s the intelligent network was the prevailing vision for the telecommunications industry. It replaced the previous vision of the 1980s, which had been the exploitation of computer technology to achieve a fully digital telephone network. As in most branches of engineering, yesterday's visions tend to become today's 'business as usual'. The use of IN principles has now become commonplace, in the same way that earlier visions, such as the 'new' digital networks of the 1980s and fibre optic transmission, also became commonplace.

However, whilst many of the original IN aims have been met, several factors conspired to make the introduction of IN a long process. These included issues such as:

◇ need for short-term financial justification
◇ vendor control of switching equipment
◇ competition with legacy switch-based features.

We now expand on the background to these issues because an appreciation of how practical realities intrude effortlessly into the implementation of a visionary concept such as IN is crucial to understanding how we can successfully evolve our networks for the future.

### 8.1.1 Financial justification

In the 1990s, the aims of intelligent networking seemed perfectly clear, and they were well articulated. The large operators needed to re-arrange their networks to enable them to introduce new service ideas quickly and cheaply and the ability to do this became the 'holy grail' in their strategic plans. Words and phrases like 'vendor-independence', 'ubiquity', 'open interfaces', 're-usable platforms', 'service independent functionality' became the currency of the newly emerging IN community, whose enthusiasm was possibly as unbounded as that of today's Internet community!

As time moved on, however, there was a realisation that good ideas needed to be turned into practical business cases, because network upgrades actually cost a lot of money. This inevitably resulted in a tempering of ideals, and a measure of pragmatism and compromise started to mellow the intelligent network expectations.

Another complication that caused delay in the arrival of IN in public networks was that it tended to fall on the first candidate IN services to justify the expense of the necessary network upgrades for the entire new IN infrastructure! Hence, the search for the 'silver bullet' service started – a single service which would make it all worthwhile, and whose backers were prepared to bear the initial brunt of the required network upgrades to move the IN vision into reality. This often proved difficult. When whole-life costs of services are analysed it quickly becomes clear that as well as the real-time network mechanisms one has to include the surrounding legacy operational support environments – the systems for billing, order handling, maintenance, etc. In fact, as is the case with most major network rearrangements, these latter costs are liable to dwarf the real-time network upgrade costs.

### 8.1.2 Vendor control

Another factor that tended to slow the speed of IN rollouts was the understandable reluctance of domestic equipment suppliers to share the operators' high degrees of enthusiasm for the idea of 'vendor-independent' networks. Operators traditionally used to buy switching equipment from a small number of vendors, who had in turn become used to the idea that much of their business was derived from selling new services through costly network upgrades, which, because of their scale, involved long lead-times and lengthy roll-out procedures.

### 8.1.3 Legacy switch-based features

Before IN, new telephony services had generally been introduced by software modifications to the switch call control software, and this usually involved a negotiated contract with the switch manufacturer. Whilst this was just the sort of lengthy process and expense that IN was intended to eliminate, ironically the IN upgrades themselves were subject to the same (or even worse) delays and expenses.

Most network operators' sales teams are under continuous pressure to introduce new services regularly in order to 'delight' – and indeed to keep – their customers in the face of the strong competition. Therefore, when a demand for a new service arose, a working implementation for it was needed as quickly as possible. In these circumstances, the network operator could be faced with a choice between a service independent IN solution that might be available in 12 months, or a vendor-tested, service-specific network upgrade providing all the functions needed, available maybe in just 6 months. With these options the obvious temptation is to choose the latter to meet existing targets.

Obviously the consequence of this 'short-term' view is of course that switch designs become more and more cluttered with disjoint functions, added in an ad hoc manner, and the risk of the features interacting with adverse consequences grows steadily and exponentially. Meanwhile the task of introducing IN-based features on these switches became more difficult because of increasingly complex call control software and erosion of potential IN business cases.

## 8.2 Tomorrow's intelligent network

At the start of the book we reported on an early authority on IN [1] who, in 1993, predicted that IN would be history by the year 2000. Accordingly, at the turn of the century we were asking whether IN was the shooting star or the discarded catalyst! We can now see that IN has in fact grown into an essential cornerstone technology in voice networks. In many ways IN has now reached a plateau of stability, and there are fundamental questions about how IN will now assist with the evolution of future networks.

An important role for IN has been the part it has played as a 'spawning-ground' for the advancement of important technologies that are now shaping the future of the industry.

An obvious example of this is the TINA project which, whilst rooted in early IN developments, quickly dissociated itself from the original 'bottom-up' IN approach in order to place new telecommunications service requirements in a wider business context. In doing so it introduced the OMG's concepts of object-orientation and distributed computing into the telecommunications arena, which have now become important concepts for next-generation distributed intelligence networks.

Another example of an IN-spawned technology is the concept of an open network API, for which pioneering work was carried out by the Parlay team. Such an API is able to effectively transform a telecommunications network into a programmable entity, allowing new service development opportunities for third party service providers. The pay-back for the network operator is that expanded service activity will lead to greater network infrastructure use, hence increasing network operator revenues. The Parlay output has now become an important standard under the auspices of the ETSI OSA.

### 8.2.1 Future services

The operation of IN within a network should obviously be entirely transparent to the users of IN services. They need know nothing about IN. IN is essentially a network re-arrangement, internal to the network operator's business, which enables advanced telephony services to be offered in an efficient and cost effective manner. Examples have been given of the type of popular IN services such as Freephone, premium rate, VPN, personal number, call screening, Calling Card, etc. In addition, the use of IN implementations of services that result from regulatory obligations, such as Number Portability and Carrier Preselection, are commonly used. This is particularly the case in the USA, as a result of which it is common for the majority of PSTN calls to use IN features. The immediate future for IN in telephony networks will undoubtedly see continued expansion and adaptation of these familiar services.

Whilst IN services contribute significantly and steadily to PSTN revenue streams in fixed networks, the fastest growth in the use of IN has recently been in mobile networks, and the popularity of the 'Pre-pay' and inter-network roaming services have been largely responsible for this. In this book we have limited our treatment of IN and mobile networks to overview explanations and some examples that illustrate the use of CAMEL in GSM in various scenarios. However, the currently most exciting prospects for the expansion of the use of IN principles is almost certainly in this field of mobile communications. For a more thorough treatment of intelligence in current and future mobile networks the reader is referred to Reference 45.

Another field where IN promises to make a significant impact is that of Internet convergence. We have described (in Chapter 6) the use of IN for managing the traffic routing for PSTN dial-up Internet access, so that it is directed to the service provider's equipment at the most appropriate point of presence. However, the biggest single issue facing the future of IN evolution is likely to be the role IN has to play in the new voice over IP (VoIP) networks. In considering the evolution of IN we have to continually bear in mind that today's IN has primarily been designed for new services on voice networks. However, the success of the Internet protocol (IP) has meant that the traditional circuit-switched networks for voice telephony may now start to give way to VoIP technology. This raises the question of the new role of IN technologies and how they transfer to the new order. In-depth coverage of the impact of network intelligence on IP networks is provided in Reference 52.

# References

Several of the references cited are for ITU, ETSI and IETF output documents that were valid at the time of publication. However, before drawing on their content it is advisable to check on their currency. This can be done at the following Internet web sites:

ITU: http/www.itu.int/publications/itut.html
ETSI: http/www.etsi.org/getastandard/home.htm
IETF: http/www.rfc.net/std1.html.

1 THORNER, J.: 'Intelligent networks' (Artech House Incorporated, 1994)
2 ATKINSON, J.: 'Telephony – a detailed exposition of the telephone exchange systems of the British Post Office, vols I & II' (Sir Isaac Pitman & Sons Ltd, 1950)
3 'CCITT signalling system No. 7', *Journal of the Institution of British Telecom Engineers* (special issue on C7), 1988, **7**, (1) Oliver, G.P. *et alia*
4 HILLS, T., and ABLETT, S.: 'Advanced intelligent networks'. Analsys Briefing Report Series No. 7, 1993
5 BROOKS, F.K.: 'The mythical man-month' (Addison Wesley, 1979)
6 ITU Recommendation Q.1200: General series, 'Intelligent network recommendation structure', Revision 1, September 1997
7 ITU Recommendation Q.763: 'Signalling system no. 7 – ISDN user part formats and codes', December 1999
8 ITU Recommendation Q.1901: 'Bearer independent call control protocol', June 2000
9 ITU Recommendation Q.1214: 'Distributed functional plane for intelligent network CS-1', October 1995
10 ETSI Standard: 'ETS 300 374-1; IN CS-1 Core INAP protocol specification, version 8' (ETSI Publications, July 1994)
11 ITU Recommendation Q.764: 'Signalling System No. 7 – ISDN user part signalling procedures', September 1997
12 ETSI Standard: 'ETS 300 403-1: DSS1 specification' (ETSI Publications, November 1998)

13  ETSI Standard: 'ETS 300 374-5: IN capability set 1 (CS-1); Core INAP; Part 5: Protocol specification at the SCF-SDF interface'
14  ITU Recommendation Q.773: 'Transaction capabilities formats and encoding', June 1997
15  WELCH, S.: 'Signalling in telecommunications networks', IEE Telecommunications Series 6 (Peter Peregrinus Ltd, 1981)
16  FLOOD, J.E. (Ed.): 'Telecommunication networks' (IEE Publications, 1997, 2nd edn)
17  HILLS, M.T.: 'Telecommunications switching principles' (George Allen & Unwin, 1984)
18  MANTERFIELD, R.: 'Common channel signalling' (Peter Peregrinus Ltd, 1991)
19  ITU Recommendation: X.200: 'Open system interconnection: the basic reference model', July 1994
20  LAW, B., and WADSWORTH, C.A.: 'SS7 signalling system no. 7: message transfer part' *Journal of the Institution of British Telecom Engineers*, April 1988
21  ITU Recommendation Q.721: 'Functional description of the signalling system no. 7: telephone user part (TUP), Blue Book, 1988
22  ETSI Standard: EN 300 195-1 v2.1.1 (2001–06) ISDN supplementary service interactions; DSS no. 1 protocol
23  CLARKE, P.G., and WADSWORTH, C.A.: 'CCITT signalling system no. 7: signalling connection control part', *British Telecommunication Engineering Journal*, 1988, **7**(1)
24  ITU Recommendation Q.711: 'Functional description of the signalling connection control part', March 2001
25  BALE, M.C.: 'Signalling in the intelligent network', in Dufour, I. (Ed.): 'Network intelligence', Chapman & Hall, 1997
26  JOHNSON, T.W., LAW, B., and ANIUS, P.: 'CCITT signalling system no. 7: transaction capabilities', *British Telecommunication Engineering Journal*, 1988, **7**(1).
27  ITU Recommendation: Q.1211: 'Introduction to intelligent network capability set 1', March 1993
28  ITU Recommendation: Q.1213: 'Global functional plane for intelligent network capability set 1', October 1995
29  TURNER, G.B.: 'Service creation', in Dufour, I. (Ed.): 'Network intelligence' Chapman & Hall, 1997
30  ITU Recommendation: Q.1215: 'Physical plane for intelligent network capability set 1', October 1995
31  ETSI/3GPP Standard: ETSI TS 101441 V7.7.0 (2001–12): Digital cellular telecommunications system (phase 2+); CAMEL phase 2, stage 2 (3GPP TS 03.78 version 7.7.0 Release 1998).
32  ITU Recommendation: Q.1224: 'Distributed functional plane for intelligent network capability set 2', September 1997
33  ITU Recommendation: Q.1222: 'Service plane for intelligent network capability set 2', September 1997
34  ITU Recommendation: Q.1223: 'Global functional plane intelligent network capability set 2', September 1997

35 UK Office of Telecommunications: 'Network Interoperability Consultation Committee: PNO-ISC Specification Number 007 (ISUP). 2001

36 The Parlay Technical Team: 'Parlay APIs 2.1 Generic call control service interfaces', March 2000. *http/www.parlay.org*

37 ETSI Standard: ES 201 915-1 V1.1.1 (2001–12) ETSI Standard – Open Service Access; Application programming interface (The Parlay Group & ETSI): DES/SPAN-120070-1. 2001

38 ETSI Standard: TR 121 905: 'Universal mobile telecommunications system (UMTS). Vocabulary for 3GPP specifications (3GPP TR 21.905)' 1999

39 TINA business model and reference points. Version 4.0 TINA-C Deliverable, May 1997. *http/www/tinac.com*

40 ITU Recommendation: Q.1204: 'Intelligent network distributed functional plane', March 1993

41 ITU Recommendation: Q.1241: 'Introduction to intelligent network capability set 4', July 2001

42 ITU Recommendation: Q.1244: 'Distributed functional plane for intelligent network capability set 4', July 2001

43 PETRACK, S., and CONROY, L.: 'IETF RFC 2848. The PINT service protocol: Extensions to SIP and SDP for IP access to telephone call services' 2000

44 FAYNBERG, I., GATO, J., LU, H., and SLUTMAN, L.: 'SPIRITS protocol requirements'. IETF Internet Draft, expires August 2002. Note that this is only referenced as work in progress at the time of publication. It is necessary to check with *www.ietf.org/internet-drafts* or *http/www.rfc.net/std1.html* to ascertain the current situation with SPIRITS work.

45 CLAPTON, A. (Ed.): 'Future mobile networks: 3G and beyond'. (IEE Books, 2001).

46 ITU Recommendation: Q.1248: 'Interface recommendation for intelligent network capability set 4' (a series of seven documents), July 2001 (pre-publication version)

47 ETSI/3GPP Standard: ETSI TS 123 078 V4.3.0 (2001–12): 'Digital cellular telecommunications system (Phase 2+) (GSM); Universal mobile telecommunications system (UMTS); Customised applications for mobile network enhanced logic (CAMEL) Phase 3 – Stage 2 (3GPP TS 23.078 version 4.3.0 Release 4)'

48 HOLMES, P.E., and STREET, N.J.: 'Operational support systems for an intelligent network' *BT Technology Journal*, 1995, **13**(2)

49 ADAMS, E., and WILLETS, K.: 'The lean communications provider: surviving the shakeout through service management excellence' (McGraw-Hill, July 1996)

50 ITU Recommendation: M.3010: 'Principles for a telecommunications management network (TMN)', February 2000

51 ITU Recommendation: M.3040 'TMN management functions', February 2000

52 SWALE, R. (Ed.): 'Voice Over IP: Systems and Solutions'. (IEE Books, 2001)

# Appendix 1: List of ITU-T IN documents

The ITU-T Q-series of recommendations covers switching and signalling in telecommunications networks. Within the Q-series, recommendations numbered Q.12.... are for intelligent network standards, and the documents between Q.13.... and Q.19.... are also IN-related. The I-series covers ISDN, so the dual Q and I numbers on some documents indicate collaborative work between the IN and ISDN working groups in ITU-T. The list below gives the dates of the documents that are in force at the time of publication of this book.

| | |
|---|---|
| Q.1200 | (09/97) General series on intelligent network recommendation structure |
| Q.1201 | (10/92) Principles of IN architecture |
| Q.1202/I.328 | (09/97) IN – service plane architecture |
| Q.1203/I.329 | (09/97) IN – global functional plane architecture |
| Q.1204 | (03/93) IN – distributed functional plane architecture |
| Q.1205 | (03/93) IN – physical plane architecture |
| Q.1208 | (09/97) General aspects of the IN application protocol |
| Q.1210 | (10/95) Q.1210-series IN recommendation structure |
| Q.1211 | (03/93) Introduction to IN CS-1 |
| Q.1213 | (10/95) Global functional plane for IN CS-1 |
| Q.1214 | (10/95) Distributed functional plane for IN CS-1 |
| Q.1215 | (10/95) Physical plane for IN CS-1 |
| Q.1218 | (10/95) Interface recommendation for IN CS-1 |
| Q.1218 Add. | (09/97) Definition for two new contexts in the SDF data model |
| Q.1219 | (04/94) IN user's guide for CS-1 |
| Q.1220 | (09/97) Q.1220-series IN CS-2 recommendation structure |
| | |
| Q.1221 | (09/97) Introduction to IN CS-2 |
| Q.1222 | (09/97) Service plane for IN CS-2 |
| Q.1223 | (09/97) Global functional plane for IN CS-2 |
| Q.1224 | (09/97) Distributed functional plane for IN CS-2 |
| Q.1225 | (09/97) Physical plane for IN CS-2 |
| Q.1228 | (09/97) Interface recommendation for IN CS-2 |
| Q.1229 | (03/99) IN user's guide for CS-2 |

| Q.1231 | (12/99) Introduction to IN CS-3 |
|---|---|
| Q.1236 | (12/99) IN CS-3 – management information model, requirements and methodology |
| Q.1237 | (06/00) Extensions to IN CS-3 in support of B-ISDN |
| Q.1238 | (*to be published*) Interface recommendation for IN CS-3 |
| Q.1238.1 | (06/00) Common aspects |
| Q.1238.2 | (06/00) Interface recommendation for IN CS-3: SCF-SSF interface |
| Q.1238.3 | (06/00) Interface recommendation for IN CS-3: SCF-SRF interface |
| Q.1238.4 | (06/00) Interface recommendation for IN CS-3: SCF-SDF interface |
| Q.1238.5 | (06/00) Interface recommendation for IN CS-3: SDF-SDF interface |
| Q.1238.6 | (06/00) Interface recommendation for IN CS-3: SCF-SCF interface |
| Q.1238.7 | (06/00) Interface recommendation for IN CS-3: SCF-CUSF interface |
| | |
| Q.1241 | (08/01)* Introduction to IN CS-4 |
| Q.1244 | (08/01)* Distributed functional plane for IN CS-4 |
| Q.1248.1 | (08/01)* Interface recommendation for IN CS-4 – Common aspects |
| Q.1248.2 | (08/01)* Interface recommendation for IN CS-4: SCF-SSF interface |
| Q.1248.3 | (08/01)* Interface recommendation for IN CS-4: SCF-SRF interface |
| Q.1248.4 | (08/01)* Interface recommendation for IN CS-4: SCF-SDF interface |
| Q.1248.5 | (08/01)* Interface recommendation for IN CS-4: SDF-SDF interface |
| Q.1248.6 | (08/01)* Interface recommendation for IN CS-4: SCF-SCF interface |
| Q.1248.7 | (08/01)* Interface recommendation for IN CS-4: SCF-CUSF interface |
| Q.1290 | (05/98) Glossary of terms used in the definition of intelligent networks |
| | |
| Q.1300 | (10/95) TASC Telecommunication applications for switches and computers. General overview |
| Q.1301 | (10/95) TASC architecture |
| Q.1302 | (10/95) TASC functional services |
| Q.1303 | (10/95) TASC management: architecture, methodology and requirements |
| Q.1400 | (02/95) Architecture framework for the development of signalling and OA&M protocols using OSI concepts |
| | |
| Q.1521 | (06/00) Requirements on underlying networks and signalling protocols to support UPT |
| Q.1531 | (06/00) UPT security requirements for Service Set 1 |
| Q.1551 | (06/97) Application of CS-1 INAP UPT Service Set 1 |
| Q.1600 | (09/97) SS7 – Interaction between ISUP and INAP |
| Q.1600bis | (12/99) SS7 – Interaction between ISUP and INAP: test suite structure |
| Q.1601 | (12/99) SS7 – Interaction between N-ISDN and INAP CS2 |
| Q.1701 | (03/99) Framework for IMT-2000 networks |
| Q.1711 | (03/99) Network functional model for IMT-2000 |
| Q.1721 | (06/00) Information flows for IMT-2000 capability set 1 |
| Q.1731 | (06/00) Radio-technology-independent requirements for IMT-2000 layer 2 radio interface |
| Q.1751 | (06/00) Inter-network signalling requirements for IMT-2000 capability set 1 |
| Q.1901 | (06/00) Bearer independent call control protocol. |

*\* These are new recommendations, approved pre-published versions available (July 2001) Information in this Appendix is derived from the ITU-T web-site (http://www.itu.int) on 18 December 2001.*

# Appendix 2: Additional CS-1 INAP operations

**Additional INAP operations specified in ITU-T CS-1**

The following is a list of the 26 operations included in ITU-T CS-1, but excluded from ETSI Core INAP and ITU-T CS-1R:

| *IN operation* |
| --- |
| analysedInformation |
| analyseInformation |
| CollectedInformation |
| CancelStatusReportRequest |
| HoldCallInNetwork |
| OAnswer |
| OCalledPartyBusy |
| ODisconnect |
| OMidCall |
| ONoAnswer |
| OriginationAttemptAuthorized |
| RequestCurrentStatusReport |
| RequestEveryStatusChangeReport |
| RequestFirstStatusMatchReport |
| RouteSelectFailure |
| SelectFacility |
| SelectRoute |
| StatusReport |
| TAnswer |
| TCalledPartyBusy |
| TDisconnect |
| TMidCall |
| TNoAnswer |
| Query |
| SdfResponse |
| UpdateData |

# Appendix 3: Example SS7 message sequence coding

SS7 SCCP and TCAP support the non-connection-oriented messaging transactions required by IN. The operation of SCCP and TCAP were outlined in Chapter 3 and we now illustrate the message coding details for a typical IN transaction. In fact the particular message flow illustrated here is for a mobile call involving a CAP (CAMEL application protocol) request. As we discussed earlier, there is strong equivalence between the CAP and the INAP protocols. There are some application parameter differences, but the behaviour of SCCP and TCAP in support of the messaging is identical in both cases and so the illustration here would be similar for an INAP transaction scenario.

We show the messaging required for a call where the IN service logic requests the SSF (the gsmSSF in this case) for end of call notification. This might be used, for instance, where the SCF needs to log the call's duration, perhaps in order to decrement the customer's prepaid account by the cost of the call.

Figure A3.1 shows the seven messages in the example sequence, and transcripts of the individual messages themselves are shown on the following pages.

The first event shown is the receipt by the gsmSCF of the CAP I_DP in a TCAP Begin message. The accompanying parameters transferred are:

◇ service key
◇ called party number
◇ calling party number
◇ calling party's category
◇ IP SSP capabilities
◇ bearer capability
◇ event type BCSM
◇ IMSI (international mobile subscriber identity)
◇ Call reference number
◇ MSC address.

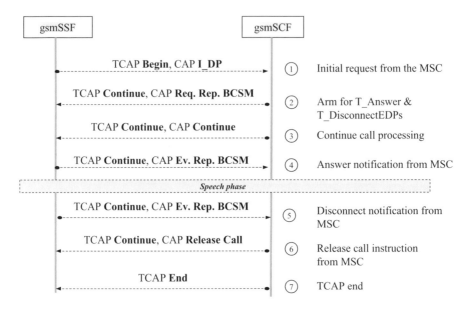

*Figure A3.1   TCAP messages illustrated*

The first seven parameters are taken directly from Core INAP (details in Chapter 2) and the last three are CAMEL-specific.

About 500 ms after the gsmSCF receives the I_DP it returns a Request Report BCSM Operation (500 ms is a typical duration for the processing on an IN request in an SCF). This is the second message of the sequence and it is a TCAP Continue message. It includes a single IN parameter that gives details of the particular events for which the service logic requires the SSP to monitor. Shortly following this is a third message, in the same direction as the first, and this is another TCAP Continue, carrying an IN Continue operation, requiring no parameters. The purpose of this message is to instruct the gsmSSP to continue with normal call processing. In this case the gsmSSP will use the digits dialled by the caller to instigate the routing of the call towards the required destination.

Eight seconds later an 'answer' notification is sent from the gsmSSF to the gsmSCF in another Continue message and the call enters the 'speech' phase. After about 38 s a 'disconnect' is generated by the first leg of the call and the gsmSSF consequently sends another TCAP Continue message carrying another Event Report BCSM operation.

The 6th message carries a Release Call operation from the gsmSCF, and this is in another TCAP Continue, so the transaction remains open until the 7th message, where the gsmSCF sends an explicit TCAP End message (with no CAP operation this time).

The following transcripts for the seven messages of this example call sequence illustrate the overheads involved at each SS7 layer. We show the actual hexadecimal code for each message followed by a decomposition and interpretation for this code. Each message has MTP and SCCP header information as well as TCAP transaction, dialogue and component headers as a prelude to the actual IN operations and parameters.

MTP headers contain the sequence numbers, the point codes and the service information octet, which holds the service indicator (value '3' for SCCP in these messages) and the network indicator (value '2' for national in these messages). SCCP message headers include the called and calling SCCP addresses with indicators giving information on the particular assemblies of global titles, subsystem numbers and point code addressing used for each message. The use of these addressing options was outlined in Chapter 3. The message sequences also show how the TCAP headers are used. Details of interpretation for TCAP are in Reference 14.

*Message No. 1:*

*CAMEL I_DP operation+parameters in a TCAP BEGIN message*
*gsmSSF* ⟶ *gsmSCF*
Time: 18:47:53,853

| HEX | 0 | 1 | 2 | 3 | 4 | 5 | 6 | 7 | 8 | 9 | A | B | C | D | E | F |
|---|---|---|---|---|---|---|---|---|---|---|---|---|---|---|---|---|
| 0 | cd | 97 | 3f | 83 | 20 | 00 | 0c | b0 | 09 | 81 | 03 | 0e | 12 | 0b | 52 | 05 |
| 10 | 00 | 12 | 04 | 21 | 43 | 65 | 87 | 19 | 32 | 04 | 43 | 30 | 00 | 05 | 7b | 62 |
| 20 | 79 | 48 | 04 | 2b | 09 | 03 | 00 | 6b | 1e | 28 | 1c | 06 | 07 | 00 | 11 | 86 |
| 30 | 05 | 01 | 01 | 01 | a0 | 11 | 60 | 0f | 80 | 02 | 07 | 80 | a1 | 09 | 06 | 07 |
| 40 | 04 | 00 | 00 | 01 | 00 | 32 | 00 | 6c | 51 | a1 | 4f | 02 | 01 | 01 | 02 | 01 |
| 50 | 00 | 30 | 47 | 80 | 02 | 00 | de | 82 | 08 | 04 | 90 | 21 | 43 | 65 | 11 | 11 |
| 60 | 11 | 83 | 07 | 83 | 13 | 88 | 88 | 11 | 11 | 01 | 85 | 01 | 0a | 88 | 01 | 04 |
| 70 | bb | 05 | 80 | 03 | 90 | 90 | a3 | 9c | 01 | 0c | 9f | 32 | 08 | 72 | 02 | 03 |
| 80 | 01 | 00 | 68 | 90 | f3 | 9f | 36 | 08 | 03 | 01 | 0c | 3e | 00 | 00 | 01 | fe |
| 90 | 9f | 37 | 07 | 22 | 22 | 22 | 22 | 22 | 22 | 22 | | | | | | |

**Translation for Message No. 1:**

| BITMASK | ID name | Comment or value |
|---|---|---|
| **MTP header information:** | | |
| -1001101 | Backward Sequence Number | 77 |
| 1------- | Backward Indicator Bit | 1 |
| -0010111 | Forward Sequence Number | 23 |
| 1------- | Forward Indicator Bit | 1 |
| --111111 | Length Indicator | 63 |
| 00------ | Spare | 0 |
| ----0011 | Service Indicator | SCCP |
| --00---- | Sub-service: Priority | Spare/priority 0 (USA only) |
| 10------ | Sub-service: Network Ind. | National message |
| **b14*** | Destination Point Code | 32 |

| BITMASK | ID name | Comment or value |
|---|---|---|
| **b14*** | Originating Point Code | 48 |
| ***SCCP message header:*** | | |
| 1011---- | Signalling Link Selection | 11 |
| 00001001 | SCCP Message Type | 9 (Unitdata) |
| ----0001 | Protocol Class | Class 1 |
| 1000---- | Message Handling | Return message on error |
| 00000011 | Pointer to parameter | 3 |
| 00001110 | Pointer to parameter | 14 |
| 00010010 | Pointer to parameter | 18 |
| ***Called address parameter:*** | | |
| 00001011 | Parameter Length | 11 |
| -------0 | Point Code Indicator | PC absent |
| ------1- | Sub-system No. Indicator | SSN present |
| --0100-- | Global Title Indicator | Has translation, n-plan, encoding scheme, nature of address |
| -1------ | Routing Indicator | Route on DPC + Sub-system No. |
| 0------- | For national use | 0 |
| 00000101 | Sub-system number | CAP |
| 00000000 | Translation Type | Not used |
| ----0010 | Encoding Scheme | BCD, even number of digits |
| 0001---- | Numbering Plan | ISDN/Telephony (E.164/E.163) |
| -0000100 | Nat. of Address Indicator | International number |
| 0------- | Spare | 0 |
| ***B6*** | Called Address Signals | 123456789123 |
| ***Calling address parameter:*** | | |
| 00000100 | Parameter Length | 4 |
| -------1 | Point Code Indicator | PC present |
| ------1- | Sub-system No. Indicator | SSN present |
| --0000-- | Global Title Indicator | No global title included |
| -1------ | Routing Indicator | Route on DPC + Sub-system No. |
| 0------- | For national use | 0 |
| **b14*** | Calling Party SPC | 48 |
| 00------ | Spare | 0 |
| 00000101 | Sub-system number | CAP |
| Data parameter | | |
| 01111011 | Parameter Length | 123 |
| **B123** | Data | 62 79 48 04 2b 09 03 00 6b 1e28 |

**TCAP transaction portion**

**Begin:**

| | | |
|---|---|---|
| 01100010 | Tag | |
| 01111001 | Length | 121 |

| BITMASK | ID name | Comment or value |
|---------|---------|------------------|
| **1 Origination Transaction ID** | | |
| 01001000 | Tag | |
| 00000100 | Length | 4 |
| ***B4*** | Orig Trans ID | 722010880 |
| **2 Dialogue Portion** | | |
| 01101011 | Tag | |
| 00011110 | Length | 30 |
| **2.1 Dialogue External** | | |
| 00101000 | Tag | |
| 00011100 | Length | 28 |
| **2.1.1 Dialogue Object ID** | | |
| 00000110 | Tag | |
| 00000111 | Length | 7 |
| 00000000 | Authority | CCITT Recommendation |
| 00010001 | Name Form | q |
| 10000110 | Rec Number | 7 |
| 00000101 | Rec Number | 73 |
| 00000001 | AS | 1 |
| 00000001 | Dialog-AS | Dialogue PDU |
| 00000001 | Version | 1 |
| **2.1.2 Dialogue Single ASN1** | | |
| 10100000 | Tag | |
| 00010001 | Length | 17 |
| **2.1.2.1 Dialogue Request** | | |
| 01100000 | Tag | |
| 00001111 | Length | 15 |
| **2.1.2.1.1 Protocol Version** | | |
| 10000000 | Tag | |
| 00000010 | Length | 2 |
| 00000111 | UnusedBits | 7 |
| 1------- | Version 1 | Yes |
| -0000000 | Filler | 0 |
| **2.1.2.1.2 Application Context Name (ACN)** | | |
| 10100001 | Tag | |
| 00001001 | Length | 9 |
| **2.1.2.1.2.1 ACN Object ID** | | |
| 00000110 | Tag | |
| 00000111 | Length | 7 |
| 0000---- | ObjId | CCITT |
| ----0100 | Organisation | Identified-organisa-tion |
| 00000000 | | ETSI |
| 00000000 | Domain | Mobile Domain |
| 00000001 | Mobile Sub-domain | GSM-Network |
| 00000000 | Common Component ID | AC-ID |
| 00110010 | Application Context | cap-gsmSSF-to-gsmSCF |
| 00000000 | Version | Version1 |
| **3 TCAP Component Portion** | | |
| 01101100 | Tag | |
| 01010001 | Length | 81 |

| BITMASK | ID name | Comment or value |
|---|---|---|
| **3.1 Invoke** | | |
| 10100001 | Tag | |
| 01001111 | Length | 79 |
| **3.1.1 Invoke ID** | | |
| 00000010 | Tag | |
| 00000001 | Length | 1 |
| 00000001 | Invoke ID Value | 1 |
| **3.1.2 CAP Local Operation** | | |
| 00000010 | Tag | |
| 00000001 | Length | 1 |
| 00000000 | Operation Code | **Initial DP** |
| **3.1.3 Parameter Sequence** | | |
| 00110000 | Tag | |
| 01000111 | Length | 71 |
| **3.1.3.1 Service Key** | | |
| 10000000 | Tag]) | |
| 00000010 | Length | 2 |
| ***B2*** | Service Key | 222 |
| **3.1.3.2 Called Party Number** | | |
| 10000010 | Tag | |
| 00001000 | Length | 8 |
| 0------- | Odd/Even Indicator | Even number of address signals |
| -0000100 | Nature of Addr. Indicator | International number |
| 1------- | Internal Netw. No. Indic. | rout.2 internal net.- not allwd |
| -001---- | Numbering Plan Indicator | ISDN numbering plan |
| ----0000 | Spare | Spare |
| ***B6*** | cld.pty.no. Address Signal | 123456111111 |
| **3.1.3.3 Calling Party Number** | | |
| 10000011 | Tag | |
| 00000111 | Length | 7 |
| 1------- | Odd/Even Indicator | Odd number of address signals |
| -0000011 | Nature of Addr. Indicator | National [significant] number |
| 0------- | Number Incomplete Indic. | Complete |
| -001---- | Numbering Plan Indicator | ISDN numbering plan |
| ----00-- | Addr. Present. Restr. Ind. | Presentation allowed |
| ------11 | Screening Indicator | Network provided |
| **b36*** | clg.pty.no. Address Signal | 888811111 |
| **3.1.3.4 Calling Party's Category** | | |
| 10000101 | Tag | |
| 00000001 | Length | 1 |
| 00001010 | CallingPartysCategory | Ordinary calling subscriber |
| **3.1.3.5 IPSSP Capabilities** | | |
| 10001000 | Tag | |
| 00000001 | Length | 1 |
| 0------- | End of Standard Part | End of standard part |
| -00----- | Reserved | 0 |

| BITMASK | ID name | Comment or value |
|---|---|---|
| ---0---- | Voice Announ.Gen.from Text | Not supported |
| ----0--- | Voice Inf.via Voice R. | Not supported |
| -----1-- | Voice Inf.via Speech R. | Supported |
| ------0- | Voice Back Y/N | Not supported |
| -------0 | IP Routing Address Y/N | Not supported |

**3.1.3.6 Bearer Capability Constr**

| | | |
|---|---|---|
| 10111011 | Tag | |
| 00000101 | Length | 5 |

**3.1.3.6.1 Bearer Cap**

| | | |
|---|---|---|
| 10000000 | Tag | |
| 00000011 | Length | 3 |
| 1------- | Extension Indicator | Fixed to last octet |
| -00----- | Coding Standard | CCITT standardised coding |
| ---10000 | Info. Trans. Cap | 3.1 kHz audio |
| 1------- | Extension Indicator | Last octet |
| -00----- | Transfer Mode | Circuit mode |
| ---10000 | Info. Transfer Rate | 64 kbit/s |
| 1------- | Extension Indicator | Last octet |
| -01----- | Layer | User Info. Layer 1 |
| ---00011 | User Info. L1 Protocol | Rec. G711, A-Law |

**3.1.3.7 Event Type BCSM**

| | | |
|---|---|---|
| 10011100 | Tag | |
| 00000001 | Length | 1 |
| 00001100 | Event Type Bcsm | **termAttemptAuthorised** |

**3.1.3.8 IMSI** (International Mobile Subscriber Identity)

| | | |
|---|---|---|
| ***B2*** | Tag | |
| 00001000 | Length | 8 |
| **b60*** | MCC + MNC + MSIN | 272030100086093 |
| 1111---- | FILLER | 15 |

**3.1.3.9 Call Reference Number**

| | | |
|---|---|---|
| ***B2*** | Tag | |
| 00001000 | Length | 8 |
| ***B8*** | Call Reference | 03 01 0c 3e 00 00 01 fe |

**3.1.3.10 MSC Address**

| | | |
|---|---|---|
| ***B2*** | Tag | (CONT P [55]) |
| 00000111 | Length | 7 |
| ***B7*** | **MSC Address** | 22 22 22 22 22 22 22 |

*Message No. 2:*
*CAMEL Request Report BCSM operation + parameters in a TCAP*
*Continue message*
gsmSSF ← gsmSCF
Time: 18:47:54,409

| HEX | 0 | 1 | 2 | 3 | 4 | 5 | 6 | 7 | 8 | 9 | A | B | C | D | E | F |
|-----|---|---|---|---|---|---|---|---|---|---|---|---|---|---|---|---|
| 0 | 98 | d0 | 3f | 83 | 30 | 00 | 08 | 90 | 09 | 01 | 03 | 07 | 0b | 04 | 43 | 30 |
| 10 | 00 | 05 | 04 | 43 | 20 | 00 | 05 | 6f | 65 | 6d | 48 | 04 | 00 | 6a | b6 | e3 |
| 20 | 49 | 04 | 2b | 09 | 03 | 00 | 6b | 2a | 28 | 28 | 06 | 07 | 00 | 11 | 86 | 05 |
| 30 | 01 | 01 | 01 | a0 | 1d | 61 | 1b | 80 | 02 | 07 | 80 | a1 | 09 | 06 | 07 | 04 |
| 40 | 00 | 00 | 01 | 00 | 32 | 00 | a2 | 03 | 02 | 01 | 00 | a3 | 05 | a1 | 03 | 02 |
| 50 | 01 | 00 | 6c | 33 | a1 | 31 | 02 | 01 | 02 | 02 | 01 | 17 | 30 | 29 | a0 | 27 |
| 60 | 30 | 0b | 80 | 01 | 0f | 81 | 01 | 01 | a2 | 03 | 80 | 01 | 02 | 30 | 0b | 80 |
| 70 | 01 | 11 | 81 | 01 | 00 | a2 | 03 | 80 | 01 | 01 | 30 | 0b | 80 | 01 | 11 | 81 |
| 80 | 01 | 00 | a2 | 03 | 80 | 01 | 02 | | | | | | | | | |

## Translation for Message No. 2:

| BITMASK | ID name | Comment or value |
|---------|---------|------------------|
| **MTP header information:** | | |
| -0011000 | Backward Sequence Number | 24 |
| 1------- | Backward Indicator Bit | 1 |
| -1010000 | Forward Sequence Number | 80 |
| 1------- | Forward Indicator Bit | 1 |
| --111111 | Length Indicator | 63 |
| 00------ | Spare | 0 |
| ----0011 | Service Indicator | SCCP |
| --00---- | Sub-service: Priority | Spare/priority 0 (USA only) |
| 10------ | Sub-service: Network Ind | National message |
| **b14*** | Destination Point Code | 48 |
| **b14*** | Originating Point Code | 32 |
| **SCCP message header:** | | |
| 1001---- | Signalling Link Selection | 9 |
| 00001001 | SCCP Message Type | 9 (Unitdata) |
| ----0001 | Protocol Class | Class 1 |
| 0000---- | Message Handling | No special options |
| 00000011 | Pointer to parameter | 3 |
| 00000111 | Pointer to parameter | 7 |
| 00001011 | Pointer to parameter | 11 |
| **Called address parameter:** | | |
| 00000100 | Parameter Length | 4 |
| -------1 | Point Code Indicator | PC present |
| ------1- | Subsystem No. Indicator | SSN present |
| --0000-- | Global Title Indicator | No global title included |

| BITMASK | ID name | Comment or value |
|---------|---------|------------------|
| -1------ | Routing Indicator | Route on DPC + Subsystem No. |
| 0------- | For national use | 0 |
| **b14*** | Called Party SPC | 48 |
| 00------ | Spare | 0 |
| 00000101 | Sub-system number | CAP |
| **Calling address parameter:** | | |
| 00000100 | Parameter Length | 4 |
| -------1 | Point Code Indicator | PC present |
| ------1- | Sub-system No. Indicator | SSN present |
| --0000-- | Global Title Indicator | No global title included |
| -1------ | Routing Indicator | Route on DPC + Subsystem No. |
| 0------- | For national use | 0 |
| **b14*** | Calling Party SPC | 32 |
| 00------ | Spare | 0 |
| 00000101 | Sub-system number | CAP |
| Data parameter | | |
| 01101111 | Parameter length | 111 |
| **B111** | Data | 65 6d 48 04 00 6a b6 e3 49 042b |

**TCAP transaction portion**
           **Continue:**

| | | |
|---------|---------|------------------|
| 01100101 | Tag | |
| 01101101 | Length | 109 |

**1 Origination Transaction ID**

| | | |
|---------|---------|------------------|
| 01001000 | Tag | |
| 00000100 | Length | 4 |
| ***B4*** | Orig Trans ID | 6993635 |

**2 Destination Transaction ID**

| | | |
|---------|---------|------------------|
| 01001001 | Tag | |
| 00000100 | Length | 4 |
| ***B4*** | Dest Trans ID | 722010880 |

**3 Dialogue Portion**

| | | |
|---------|---------|------------------|
| 01101011 | Tag | |
| 00101010 | Length | 42 |

**3.1 Dialogue External**

| | | |
|---------|---------|------------------|
| 00101000 | Tag | |
| 00101000 | Length | 40 |

**3.1.1 Dialogue Object ID**

| | | |
|---------|---------|------------------|
| 00000110 | Tag | |
| 00000111 | Length | 7 |
| 00000000 | Authority | CCITT Recommendation |
| 00010001 | Name Form | q |
| 10000110 | Rec Number | 7 |
| 00000101 | Rec Number | 73 |
| 00000001 | AS | 1 |
| 00000001 | Dialog-AS | Dialogue PDU |
| 00000001 | Version | 1 |

| BITMASK | ID name | Comment or value |
|---|---|---|
| **3.1.2 Dialogue single ASN1** | | |
| 10100000 | Tag | |
| 00011101 | Length | 29 |
| **3.1.2.1 Dialogue Response** | | |
| 01100001 | Tag | |
| 00011011 | Length | 27 |
| **3.1.2.1.1 Protocol Version** | | |
| 10000000 | Tag | |
| 00000010 | Length | 2 |
| 00000111 | UnusedBits | 7 |
| 1------- | Version 1 | Yes |
| -0000000 | Filler | 0 |
| **3.1.2.1.2 Application Context Name** | | |
| 10100001 | Tag | |
| 00001001 | Length | 9 |
| **3.1.2.1.2.1 ACN Object ID** | | |
| 00000110 | Tag | (UNIV P Obj Identifier) |
| 00000111 | Length | 7 |
| 0000---- | ObjId | CCITT |
| ----0100 | Organisation | Identified-organisation |
| 00000000 | | ETSI |
| 00000000 | Domain | Mobile Domain |
| 00000001 | Mobile Sub-domain | GSM-Network |
| 00000000 | Common Component ID | AC-ID |
| 00110010 | Application Context | cap-gsmSSF-to-gsmSCF |
| 00000000 | Version | Version1 |
| **3.1.2.1.3 Result** | | |
| 10100010 | Tag | |
| 00000011 | Length | 3 |
| **3.1.2.1.3.1 Associate Result** | | |
| 00000010 | Tag | |
| 00000001 | Length | 1 |
| 00000000 | Associate Result | Accepted |
| **3.1.2.1.4 Result Source Diagnostic** | | |
| 10100011 | Tag | |
| 00000101 | Length | 5 |
| **3.1.2.1.4.1 Dialogue Service User** | | |
| 10100001 | Tag | |
| 00000011 | Length | 3 |
| **3.1.2.1.4.1.1 Dialogue Service User Value** | | |
| 00000010 | Tag | |
| 00000001 | Length | 1 |
| 00000000 | Dialogue Service User | Null |
| **4 TCAP Component Portion** | | |
| 01101100 | Tag | |
| 00110011 | Length | 51 |
| **4.1 Invoke** | | |
| 10100001 | Tag | |

| BITMASK | ID name | Comment or value |
|---|---|---|
| 00110001 | Length | 49 |

**4.1.1 Invoke ID**

| | | |
|---|---|---|
| 00000010 | Tag | |
| 00000001 | Length | 1 |
| 00000010 | Invoke ID value | 2 |

**4.1.2 CAP Local Operation**

| | | |
|---|---|---|
| 00000010 | Tag | |
| 00000001 | Length | 1 |
| 00010111 | Operation Code | **Request Report BCSME-vent** |

**4.1.3 Parameter Sequence**

| | | |
|---|---|---|
| 00110000 | Tag | |
| 00101001 | Length | 41 |

**4.1.3.1 BCSM Events**

| | | |
|---|---|---|
| 10100000 | Tag | |
| 00100111 | Length | 39 |

**4.1.3.1.1 BCSM Event**

| | | |
|---|---|---|
| 00110000 | Tag | |
| 00001011 | Length | 11 |

**4.1.3.1.1.1 Event Type BCSM**

| | | |
|---|---|---|
| 10000000 | Tag | |
| 00000001 | Length | 1 |
| 00001111 | Event Type Bcsm | **tAnswer** |

**4.1.3.1.1.2 Monitor Mode**

| | | |
|---|---|---|
| 10000001 | Tag | |
| 00000001 | Length | 1 |
| 00000001 | Monitor Mode | **Notify and continue** |

**4.1.3.1.1.3 Leg Id Constr**

| | | |
|---|---|---|
| 10100010 | Tag | |
| 00000011 | Length | 3 |

**4.1.3.1.1.3.1 Sending Side ID**

| | | |
|---|---|---|
| 10000000 | Tag | |
| 00000001 | Length | 1 |
| 00000010 | Sending Side Id | **Leg2** |

**4.1.3.1.2 BCSM Event**

| | | |
|---|---|---|
| 00110000 | Tag | |
| 00001011 | Length | 11 |

**4.1.3.1.2.1 Event Type BCSM**

| | | |
|---|---|---|
| 10000000 | Tag | |
| 00000001 | Length | 1 |
| 00010001 | Event Type Bcsm | **tDisconnect** |

**4.1.3.1.2.2 Monitor Mode**

| | | |
|---|---|---|
| 10000001 | Tag | |
| 00000001 | Length | 1 |
| 00000000 | Monitor Mode | **Interrupted** |

**4.1.3.1.2.3 Leg Id Constr**

| | | |
|---|---|---|
| 10100010 | Tag | |
| 00000011 | Length | 3 |

**4.1.3.1.2.3.1 Sending Side ID**

| | | |
|---|---|---|
| 10000000 | Tag | |
| 00000001 | Length | 1 |

| BITMASK | ID name | Comment or value |
|---------|---------|------------------|
| 00000001 | Sending Side Id | **Leg1** |
| **4.1.3.1.3 BCSM Event** | | |
| 00110000 | Tag | |
| 00001011 | Length | 11 |
| **4.1.3.1.3.1 Event Type BCSM** | | |
| 10000000 | Tag | |
| 00000001 | Length | 1 |
| 00010001 | Event Type Bcsm | **tDisconnect** |
| **4.1.3.1.3.2 Monitor Mode** | | |
| 10000001 | Tag | |
| 00000001 | Length | 1 |
| 00000000 | Monitor Mode | Interrupted |
| **4.1.3.1.3.3 Leg Id Constr** | | |
| 10100010 | Tag | |
| 00000011 | Length | 3 |
| **4.1.3.1.3.3.1 Sending Side ID** | | |
| 10000000 | Tag | |
| 00000001 | Length | 1 |
| 00000010 | Sending Side Id | **Leg2** |

## Message No. 3:

### CAMEL Continue operation in a TCAP Continue message
### gsmSSF ⟵ gsmSCF
Time: 18:47:54,416

| HEX | 0 | 1 | 2 | 3 | 4 | 5 | 6 | 7 | 8 | 9 | A | B | C | D | E | F |
|-----|---|---|---|---|---|---|---|---|---|---|---|---|---|---|---|---|
| 0   | 98 | d1 | 2d | 83 | 30 | 00 | 08 | 90 | 09 | 01 | 03 | 07 | 0b | 04 | 43 | 30 |
| 10  | 00 | 05 | 04 | 43 | 20 | 00 | 05 | 18 | 65 | 16 | 48 | 04 | 00 | 6a | b6 | e3 |
| 20  | 49 | 04 | 2b | 09 | 03 | 00 | 6c | 08 | a1 | 06 | 02 | 01 | 03 | 02 | 01 | 1f |

## Translation for Message No. 3:

| BITMASK | ID name | Comment or value |
|---------|---------|------------------|
| **MTP header information:** | | |
| -0011000 | Backward Sequence Number | 24 |
| 1------- | Backward Indicator Bit | 1 |
| -1010001 | Forward Sequence Number | 81 |
| 1------- | Forward Indicator Bit | 1 |
| --101101 | Length Indicator | 45 |
| 00------ | Spare | 0 |
| ----0011 | Service Indicator | SCCP |
| --00---- | Sub-service: Priority | Spare/priority 0 (USA only) |
| 10------ | Sub-service: Network Ind. | National message |
| **b14*** | Destination Point Code | 48 |
| **b14*** | Originating Point Code | 32 |
| **SCCP message header:** | | |
| 1001---- | Signalling Link Selection | 9 |
| 00001001 | SCCP Message Type | 9 (UDT) |
| ----0001 | Protocol Class | Class 1 |
| 0000---- | Message Handling | No special options |
| 00000011 | Pointer to parameter | 3 |
| 00000111 | Pointer to parameter | 7 |
| 00001011 | Pointer to parameter | 11 |
| **Called address parameter:** | | |
| 00000100 | Parameter Length | 4 |
| -------1 | Point Code Indicator | PC present |
| ------1- | Sub-system No. Indicator | SSN present |
| --0000-- | Global Title Indicator | No global title included |
| -1------ | Routing Indicator | Route on DPC + Sub-system No. |
| 0------- | For national use | 0 |
| **b14*** | Called Party SPC | 48 |
| 00------ | Spare | 0 |
| 00000101 | Sub-system number | CAP |
| **Calling address parameter:** | | |
| 00000100 | Parameter Length | 4 |

| BITMASK | ID name | Comment or value |
|---------|---------|------------------|
| -------1 | Point Code Indicator | PC present |
| ------1- | Sub-system No. Indicator | SSN present |
| --0000-- | Global Title Indicator | No global title included |
| -1------ | Routing Indicator | Route on DPC + Sub-system No. |
| 0------- | For national use | 0 |
| **b14*** | Calling Party SPC | 32 |
| 00------ | Spare | 0 |
| 00000101 | Sub-system number | CAP |
| Data parameter | | |
| 00011000 | Parameter length | 24 |
| **B24*** | Data | 65 16 48 04 00 6a b6 e3 49 042b |

**TCAP transaction portion**
Continue:

| | | |
|---------|---------|------------------|
| 01100101 | Tag | |
| 00010110 | Length | 22 |
| **1 Origination Transaction ID** | | |
| 01001000 | Tag | |
| 00000100 | Length | 4 |
| ***B4*** | Orig Trans ID | 6993635 |
| **2 Destination Transaction ID** | | |
| 01001001 | Tag | |
| 00000100 | Length | 4 |
| ***B4*** | Dest Trans ID | 722010880 |
| **3 TCAP Component Portion** | | |
| 01101100 | Tag | |
| 00001000 | Length | 8 |
| **3.1 Invoke** | | |
| 10100001 | Tag | |
| 00000110 | Length | 6 |
| **3.1.1 Invoke ID** | | |
| 00000010 | Tag | |
| 00000001 | Length | 1 |
| 00000011 | Invoke ID value | 3 |
| **3.1.2 CAP Local Operation** | | |
| 00000010 | Tag | |
| 00000001 | Length | 1 |
| 00011111 | Operation Code | **Continue** |

## Message No. 4:

*CAMEL Event Report BCSM operation+parameters in a TCAP*
*Continue message*

gsmSSF ⟶ gsmSCF

time: 18:48:02,562

| HEX | 0 | 1 | 2 | 3 | 4 | 5 | 6 | 7 | 8 | 9 | A | B | C | D | E | F |
|-----|----|----|----|----|----|----|----|----|----|----|----|----|----|----|----|----|
| 0 | f2 | b2 | 38 | 83 | 20 | 00 | 0c | b0 | 09 | 81 | 03 | 05 | 07 | 02 | 42 | 05 |
| 10 | 02 | 42 | 05 | 27 | 65 | 25 | 48 | 04 | 2b | 09 | 03 | 00 | 49 | 04 | 00 | 6a |
| 20 | b6 | e3 | 6c | 17 | a1 | 15 | 02 | 01 | 02 | 02 | 01 | 18 | 30 | 0d | 80 | 01 |
| 30 | 0f | a3 | 03 | 81 | 01 | 02 | a4 | 03 | 80 | 01 | 01 | | | | | |

## Translation for Message No. 4:

| BITMASK | ID name | Comment or value |
|---------|---------|------------------|
| **MTP header information:** | | |
| -1110010 | Backward Sequence Number | 114 |
| 1------- | Backward Indicator Bit | 1 |
| -0110010 | Forward Sequence Number | 50 |
| 1------- | Forward Indicator Bit | 1 |
| --111000 | Length Indicator | 56 |
| 00------ | Spare | 0 |
| ----0011 | Service Indicator | SCCP |
| --00---- | Sub-service: Priority | Spare/priority 0 (USA only) |
| 10------ | Sub-service: Network Ind. | National message |
| **b14*** | Destination Point Code | 32 |
| **b14*** | Originating Point Code | 48 |
| **SCCP message header:** | | |
| 1011---- | Signalling Link Selection | 11 |
| 00001001 | SCCP Message Type | 9 |
| ----0001 | Protocol Class | Class 1 |
| 1000---- | Message Handling | Return message on error |
| 00000011 | Pointer to parameter | 3 |
| 00000101 | Pointer to parameter | 5 |
| 00000111 | Pointer to parameter | 7 |
| **Called address parameter:** | | |
| 00000010 | Parameter Length | 2 |
| -------0 | Point Code Indicator | PC absent |
| ------1- | Sub-system No. Indicator | SSN present |
| --0000-- | Global Title Indicator | No global title included |
| -1------ | Routing Indicator | Route on DPC+Sub-system No. |
| 0------- | For national use | 0 |
| 00000101 | Sub-system number | CAP |

| BITMASK | ID name | Comment or value |
|---------|---------|------------------|
| | **Calling address parameter:** | |
| 00000010 | Parameter Length | 2 |
| -------0 | Point Code Indicator | PC absent |
| ------1- | Sub-system No. Indicator | SSN present |
| --0000-- | Global Title Indicator | No global title included |
| -1------ | Routing Indicator | Route on DPC + Sub-system No. |
| 0------- | For national use | 0 |
| 00000101 | Sub-system number | CAP |
| Data parameter | | |
| 00100111 | Parameter Length | 39 |
| **B39*** | Data | 65 25 48 04 2b 09 03 00 49 0400 |
| **TCAP transaction portion** | | |
| | **Continue:** | |
| 01100101 | Tag | |
| 00100101 | Length | 37 |
| 1 Origination Transaction ID | | |
| 01001000 | Tag | |
| 00000100 | Length | 4 |
| ***B4*** | Orig Trans ID | 722010880 |
| **2 Destination Transaction ID** | | |
| 01001001 | Tag | |
| 00000100 | Length | 4 |
| ***B4*** | Dest Trans ID | 6993635 |
| **3 TCAP Component Portion** | | |
| 01101100 | Tag | |
| 00010111 | Length | 23 |
| **3.1 Invoke** | | |
| 10100001 | Tag | |
| 00010101 | Length | 21 |
| **3.1.1 Invoke ID** | | |
| 00000010 | Tag | |
| 00000001 | Length | 1 |
| 00000010 | Invoke ID value | 2 |
| **3.1.2 Local Operation** | | |
| 00000010 | Tag | |
| 00000001 | Length | 1 |
| 00011000 | Operation Code | **Event Report BCSM** |
| **3.1.3 Parameter Sequence** | | |
| 00110000 | Tag | (UNIV C sequence (of)) |
| 00001101 | Length | 13 |
| **3.1.3.1 Event Type BCSM** | | |
| 10000000 | Tag | |
| 00000001 | Length | 1 |
| 00001111 | Event Type Bcsm | **tAnswer** |
| **3.1.3.2 Leg Id Constr** | | |
| 10100011 | Tag | |
| 00000011 | Length | 3 |

| *BITMASK* | *ID name* | *Comment or value* |
|---|---|---|
| **3.1.3.2.1 Receiving Side Id** | | |
| 10000001 | Tag | |
| 00000001 | Length | 1 |
| 00000010 | Receiving Side Id | **Leg2** |
| **3.1.3.3 Misc Call Info** | | |
| 10100100 | Tag | |
| 00000011 | Length | 3 |
| **3.1.3.3.1 Message Type** | | |
| 10000000 | Tag | |
| 00000001 | Length | 1 |
| 00000001 | Message Type | Notification |

**Message No. 5:**

*CAP Event Report BCSM operation + parameters in a TCAP Continue message*

gsmSSF ⟶ gsmSCF

18:48:40,486

| HEX | 0 | 1 | 2 | 3 | 4 | 5 | 6 | 7 | 8 | 9 | A | B | C | D | E | F |
|-----|---|---|---|---|---|---|---|---|---|---|---|---|---|---|---|---|
| 0 | 85 | a9 | 3f | 83 | 20 | 00 | 0c | b0 | 09 | 81 | 03 | 05 | 07 | 02 | 42 | 05 |
| 10 | 02 | 42 | 05 | 2f | 65 | 2d | 48 | 04 | 2b | 09 | 03 | 00 | 49 | 04 | 00 | 6a |
| 20 | b6 | e3 | 6c | 1f | a1 | 1d | 02 | 01 | 03 | 02 | 01 | 18 | 30 | 15 | 80 | 01 |
| 30 | 11 | a2 | 06 | ac | 04 | 80 | 02 | 80 | 90 | a3 | 03 | 81 | 01 | 01 | a4 | 03 |
| 40 | 80 | 01 | 00 | | | | | | | | | | | | | |

## Translation for Message No. 5:

| BITMASK | ID name | Comment or value |
|---------|---------|------------------|
| **MTP header information:** | | |
| -0000101 | Backward Sequence Number | 5 |
| 1------- | Backward Indicator Bit | 1 |
| -0101001 | Forward Sequence Number | 41 |
| 1------- | Forward Indicator Bit | 1 |
| --111111 | Length Indicator | 63 |
| 00------ | Spare | 0 |
| ----0011 | Service Indicator | SCCP |
| --00---- | Sub-service: Priority | Spare/priority 0 (USA only) |
| 10------ | Sub-service: Network Ind | National message |
| **b14*** | Destination Point Code | 32 |
| **b14*** | Originating Point Code | 48 |
| **SCCP message header:** | | |
| 1011---- | Signalling Link Selection | 11 |
| 00001001 | SCCP Message Type | 9 |
| ----0001 | Protocol Class | Class 1 |
| 1000---- | Message Handling | Return message on error |
| 00000011 | Pointer to parameter | 3 |
| 00000101 | Pointer to parameter | 5 |
| 00000111 | Pointer to parameter | 7 |
| | **Called address parameter:** | |
| 00000010 | Parameter Length | 2 |
| -------0 | Point Code Indicator | PC absent |
| ------1- | Sub-system No. Indicator | SSN present |
| --0000-- | Global Title Indicator | No global title included |
| -1------ | Routing Indicator | Route on DPC + Sub-system No. |
| 0------- | For national use | 0 |
| 00000101 | Sub-system number | CAP |

| BITMASK | ID name | Comment or value |
|---------|---------|------------------|
| | **Called address parameter:** | |
| 00000010 | Parameter Length | 2 |
| -------0 | Point Code Indicator | PC absent |
| ------1- | Sub-system No. Indicator | SSN present |
| --0000-- | Global Title Indicator | No global title included |
| -1------ | Routing Indicator | Route on DPC + Sub-system No. |
| 0------- | For national use | 0 |
| 00000101 | Sub-system number | CAP |
| Data parameter | | |
| 00101111 | Parameter length | 47 |
| **B47*** | Data | 65 2d 48 04 2b 09 03 00 49 0400 |

**TCAP transaction portion**
**Continue:**

| | | |
|---------|---------|------------------|
| 01100101 | Tag | |
| 00101101 | Length | 45 |
| **1 Origination Transaction ID** | | |
| 01001000 | Tag | |
| 00000100 | Length | 4 |
| ***B4*** | Orig Trans ID | 722010880 |
| **2 Destination Transaction ID** | | |
| 01001001 | Tag | |
| 00000100 | Length | 4 |
| ***B4*** | Dest Trans ID | 6993635 |
| **3 Component Portion** | | |
| 01101100 | Tag | |
| 00011111 | Length | 31 |
| **3.1 Invoke** | | |
| 10100001 | Tag | |
| 00011101 | Length | 29 |
| **3.1.1 Invoke ID** | | |
| 00000010 | Tag | |
| 00000001 | Length | 1 |
| 00000011 | Invoke ID Value | 3 |
| **3.1.2 Local Operation** | | |
| 00000010 | Tag | |
| 00000001 | Length | 1 |
| 00011000 | Operation Code | **Event Report BCSM** |
| **3.1.3 Parameter Sequence** | | |
| 00110000 | Tag | |
| 00010101 | Length | 21 |
| **3.1.3.1 Event Type BCSM** | | |
| 10000000 | Tag | |
| 00000001 | Length | 1 |
| 00010001 | Event Type Bcsm | **tDisconnect** |
| **3.1.3.2 Event Specific Info Bcsm** | | |
| 10100010 | Tag | |
| 00000110 | Length | 6 |

| BITMASK | ID name | Comment or value |
|---|---|---|
| **3.1.3.2.1 T Disconnect Specific Info** | | |
| 10101100 | Tag | |
| 00000100 | Length | 4 |
| **3.1.3.2.1.1 Release Cause** | | |
| 10000000 | Tag | |
| 00000010 | Length | 2 |
| 1------- | Extension Indication 1 | Last octet |
| -00----- | Coding Standard | CCITT standard |
| ---0---- | Spare | Spare |
| ----0000 | Location | User |
| 1------- | Extension Indication 2 | Last octet |
| -0010000 | Cause Value | Normal call clearing |
| **3.1.3.3 Leg Id Constr** | | |
| 10100011 | Tag | |
| 00000011 | Length | 3 |
| **3.1.3.3.1 Receiving Side Id** | | |
| 10000001 | Tag | |
| 00000001 | Length | 1 |
| 00000001 | Receiving Side Id | **Leg1** |
| **3.1.3.4 Misc Call Info** | | |
| 10100100 | Tag | |
| 00000011 | Length | 3 |
| **3.1.3.4.1 Message Type** | | |
| 10000000 | Tag | |
| 00000001 | Length | 1 |
| 00000000 | Message Type | Request |

*Message No. 6:*

*CAP Release Call Operation in a TCAP Continue message*

gsmSSF ← gsmSCF

Time: 18:48:40,735

| HEX | 0 | 1 | 2 | 3 | 4 | 5 | 6 | 7 | 8 | 9 | A | B | C | D | E | F |
|-----|----|----|----|----|----|----|----|----|----|----|----|----|----|----|----|----|
| 0 | ac | 86 | 31 | 83 | 30 | 00 | 08 | 90 | 09 | 01 | 03 | 07 | 0b | 04 | 43 | 30 |
| 10 | 00 | 05 | 04 | 43 | 20 | 00 | 05 | 1c | 65 | 1a | 48 | 04 | 00 | 6a | b6 | e3 |
| 20 | 49 | 04 | 2b | 09 | 03 | 00 | 6c | 0c | a1 | 0a | 02 | 01 | 04 | 02 | 01 | 16 |
| 30 | 04 | 02 | 80 | 90 | | | | | | | | | | | | |

**Translation for Message No. 6:**

| BITMASK | ID name | Comment or value |
|---------|---------|------------------|
| *MTP header information:* | | |
| -0101100 | Backward Sequence Number | 44 |
| 1------- | Backward Indicator Bit | 1 |
| -0000110 | Forward Sequence Number | 6 |
| 1------- | Forward Indicator Bit | 1 |
| --110001 | Length Indicator | 49 |
| 00------ | Spare | 0 |
| ----0011 | Service Indicator | SCCP |
| --00---- | Sub-service: Priority | Spare/priority 0 (USA only) |
| 10------ | Sub-service: Network Ind | National message |
| **b14*** | Destination Point Code | 48 |
| **b14*** | Originating Point Code | 32 |
| *SCCP message header:* | | |
| 1001---- | Signalling Link Selection | 9 |
| 00001001 | SCCP Message Type | 9 |
| ----0001 | Protocol Class | Class 1 |
| 0000---- | Message Handling | No special options |
| 00000011 | Pointer to parameter | 3 |
| 00000111 | Pointer to parameter | 7 |
| 00001011 | Pointer to parameter | 11 |
| | **Called address parameter:** | |
| 00000100 | Parameter Length | 4 |
| -------1 | Point Code Indicator | PC present |
| ------1- | Sub-system No. Indicator | SSN present |
| --0000-- | Global Title Indicator | No global title included |
| -1------ | Routing Indicator | Route on DPC + Sub-system No. |
| 0------- | For national use | 0 |
| **b14*** | Called Party SPC | 48 |
| 00------ | Spare | 0 |
| 00000101 | Sub-system number | CAP |
| | **Calling address parameter:** | |

| BITMASK | ID name | Comment or value |
|---|---|---|
| 00000100 | Parameter Length | 4 |
| -------1 | Point Code Indicator | PC present |
| ------1- | Sub-system No. Indicator | SSN present |
| --0000-- | Global Title Indicator | No global title included |
| -1------ | Routing Indicator | Route on DPC + Sub-system No. |
| 0------- | For national use | 0 |
| **b14*** | Calling Party SPC | 32 |
| 00------ | Spare | 0 |
| 00000101 | Sub-system number | CAP |
| Data parameter |  |  |
| 00011100 | Parameter length | 28 |
| **B28*** | Data | 65 1a 48 04 00 6a b6 e3 49 042b |

**TCAP transaction portion**
**Continue:**

| 01100101 | Tag |  |
|---|---|---|
| 00011010 | Length | 26 |

**1 Origination Transaction ID**

| 01001000 | Tag |  |
|---|---|---|
| 00000100 | Length | 4 |
| ***B4*** | Orig Trans ID | 6993635 |

**2 Destination Transaction ID**

| 01001001 | Tag |  |
|---|---|---|
| 00000100 | Length | 4 |
| ***B4*** | Dest Trans ID | 722010880 |

**3 Component Portion**

| 01101100 | Tag |  |
|---|---|---|
| 00001100 | Length | 12 |

**3.1 Invoke**

| 10100001 | Tag |  |
|---|---|---|
| 00001010 | Length | 10 |

**3.1.1 Invoke ID**

| 00000010 | Tag |  |
|---|---|---|
| 00000001 | Length | 1 |
| 00000100 | Invoke ID Value | 4 |

**3.1.2 Local Operation**

| 00000010 | Tag |  |
|---|---|---|
| 00000001 | Length | 1 |
| 00010110 | Operation Code | **Release Call** |

**3.1.3 Release Call Arg**

| 00000100 | Tag |  |
|---|---|---|
| 00000010 | Length | 2 |
| 1------- | Extension Indication 1 | Last octet |
| -00----- | Coding Standard | CCITT standard |
| ---0---- | Spare | Spare |
| ----0000 | Location | User |
| 1------- | Extension Indication 2 | Last octet |
| -0010000 | Cause Value | Normal call clearing |

## Message No. 7
### TCAP End message
### gsmSSF ⟵ gsmSCF
time: 18:48:40,740

| HEX | 0 | 1 | 2 | 3 | 4 | 5 | 6 | 7 | 8 | 9 | A | B | C | D | E | F |
|-----|----|----|----|----|----|----|----|----|----|----|----|----|----|----|----|----|
| 0 | ac | 87 | 1d | 83 | 30 | 00 | 08 | 90 | 09 | 01 | 03 | 07 | 0b | 04 | 43 | 30 |
| 10 | 00 | 05 | 04 | 43 | 20 | 00 | 05 | 08 | 64 | 06 | 49 | 04 | 2b | 09 | 03 | 00 |

## Translation for Message No. 7:

| BITMASK | ID name | Comment or value |
|---------|---------|------------------|
| **MTP header information:** | | |
| -0101100 | Backward Sequence Number | 44 |
| 1------- | Backward Indicator Bit | 1 |
| -0000111 | Forward Sequence Number | 7 |
| 1------- | Forward Indicator Bit | 1 |
| --011101 | Length Indicator | 29 |
| 00------ | Spare | 0 |
| ----0011 | Service Indicator | SCCP |
| --00---- | Sub-service: Priority | Spare/priority 0 (USA only) |
| 10------ | Sub-service: Network Ind | National message |
| **b14*** | Destination Point Code | 48 |
| **b14*** | Originating Point Code | 32 |
| **SCCP message header:** | | |
| 1001---- | Signalling Link Selection | 9 |
| 00001001 | SCCP Message Type | 9 |
| ----0001 | Protocol Class | Class 1 |
| 0000---- | Message Handling | No special options |
| 00000011 | Pointer to parameter | 3 |
| 00000111 | Pointer to parameter | 7 |
| 00001011 | Pointer to parameter | 11 |
| | **Called address parameter:** | |
| 00000100 | Parameter Length | 4 |
| -------1 | Point Code Indicator | PC present |
| ------1- | Sub-system No. Indicator | SSN present |
| --0000-- | Global Title Indicator | No global title included |
| -1------ | Routing Indicator | Route on DPC + Sub-system No. |
| 0------- | For national use | 0 |
| **b14*** | Called Party SPC | 48 |
| 00------ | Spare | 0 |
| 00000101 | Sub-system number | CAP |
| | **Calling address parameter:** | |
| 00000100 | Parameter Length | 4 |
| -------1 | Point Code Indicator | PC present |
| ------1- | Sub-system No. Indicator | SSN present |

| BITMASK | ID name | Comment or value |
|---------|---------|------------------|
| --0000-- | Global Title Indicator | No global title included |
| -1------ | Routing Indicator | Route on DPC + Subsystem No. |
| 0------- | For national use | 0 |
| **b14*** | Calling Party SPC | 32 |
| 00------ | Spare | 0 |
| 00000101 | Sub-system number | CAP |
| Data parameter | | |
| 00001000 | Parameter length | 8 |
| ***B8*** | Data | 64 06 49 04 2b 09 03 00 |
| **TCAP transaction portion** | | |
| | **End:** | |
| 01100100 | Tag | |
| 00000110 | Length | 6 |
| **Destination Transaction ID** | | |
| 01001001 | Tag | |
| 00000100 | Length | 4 |
| ***B4*** | Dest Trans ID | 722010880 |

# Glossary

**3G**
Third generation of mobile communications systems, based on broadband technology. 3G follows 2G, which refers to mobile systems such as GSM, using digital speech-coding technology, and 1G (earlier analogue mobile telephony systems). 2.5G refers to intermediate technologies between 2G and 3G such as GPRS-based GSM. An example of a 3G system is ETSI's UMTS.

**3GPP**
3G Partnership Project. This was originally pioneered by ETSI and it focuses on GSM evolution to 3G. Earlier ETSI work for GSM was integrated into 3GPP to achieve a single UMTS development stream. 3GPP provides the currently prevailing worldwide standard for the core network architecture and protocols, although the radio interface standard used is subject to regional differences because of spectrum allocations and licensing variations. See also 'UMTS' below.

**ACM**
Address Complete Message. This is an ISUP (and TUP) message, sent by a destination node and relayed back to the call origination node, in reply to an IAM (or IFAM). It indicates that the address has been checked and found to be valid.

**ACP**
Action Control Point, the term used for a network switch in the early DSDC number translation system offered by AT&T in the USA.

**AD**
Adjunct – this is a CS-1 physical (network) entity that is similar to an SCP, but is more closely coupled (using faster signalling than SS7) to the SSP.

**AIN**
Advanced Intelligent Network. This term was coined in the early 1990s by Bellcore (now Telcordia) for a set of widely accepted IN standards for the USA.

**ANS**
'Answer' message. This is an ISUP (and TUP) message, sent by a destination node and relayed back to the call origination node, when the termination device identified by the previously sent address digits has answered the call. This may be automatic, in the case of a modem or ISDN device, or not, in the case of a recipient of a voice call lifting the handset.

| | |
|---|---|
| **ANSI** | American National Standards Institute. ANSI facilitates the development of US domestic standards through consensus processes and communicates US standards to the international standards arena. |
| **ANSI-41** | An abbreviation of ANSI/EIA/TIA-41, which is the joint ANSI & EIA (Electronic Industries Alliance) for mobile networking. ANSI-41 is applicable to non-GSM networks. |
| **APDU** | Application Protocol Data Unit. |
| **API** | Application Programming Interface. |
| **ARI** | AssistRequestInstructions – an INAP operation. INAP operations are described in Section 2.5.1. |
| **ASF** | ActivateServiceFiltering – an INAP operation. INAP operations are described in Section 2.5.1. |
| **ASN/1** | Abstract Syntax Notation No. 1. This is the formal message-description language used by the ITU-T for defining IN information exchanges. *http:/ /asn1.elibel.tm.fr/ refers.* |
| **ATM** | Asynchronous Transfer Mode. ATM is a flexible data transfer mechanism, based on 53-octet cells, chosen by the ITU-T as the data transport technology to be used for ITU-T broadband ISDN. |
| **AT&T** | American Telephone and Telegraph company. |
| **BCM** | Basic Call Manager. The BCM encompasses the CCF and the IN triggering part of the SSF in an IN SSP node. |
| **BCP** | Basic Call Process. BCP is an ITU-T IN SIB that represents the underlying thread of basic call control in IN service creation systems. Other SIBs are introduced to model particular service behaviour patterns and these extra SIBs are superimposed on the BCP. |
| **BCSM** | Basic Call State Machine. This is the basic IN call model, which is fundamental to ITU-T IN architecture. |
| **BCUSM** | Basic Call Unrelated State Model. This is a separate (from the BCSM) call model introduced in ITU-T CS-2 to handle non-call-related features such as location updating. *See also CUSF.* |
| **Bellcore** | Bell Communications Research – the previous name for the US Telcordia Technologies Incorporated, which develops network solutions, including IN architectures. |
| **BICC** | Bearer-Independent Call Control. |
| **BSC** | The Base Station Controller in a mobile network. |
| **BSS** | Base Station System, in a mobile network. The BSS contains the BTS and BSC functions. |
| **BSSMAP** | Base Station System Mobile Application Part of the SS7 signalling system. |
| **BTNR** | British Telecom Network Requirements. Early 'de facto' scheme of documenting British Telecom's UK network interworking standards. |
| **BTS** | The Base Transceiver Station in a mobile network. |
| **C7** | ITU-T (CCITT) Signalling System No. 7 – usually called 'SS7' nowadays. |

| CAMEL | Customised Application of Mobile networks Enhanced Logic. CAMEL an intelligent networking protocol, like INAP, but optimised for GSM networks. CAMEL is the IN standard for GSM based networks and WIN is the IN standard for non-GSM mobile networks. |
|---|---|
| CAP | CAMEL Application Protocol. This is the SS7 IN signalling protocol for CAMEL. It is closely related to INAP. |
| CCAF | Call Control Agent Function. This is an IN functional entity, in the ITU-T distributed functional plane, which represents a user who has more sophisticated network interface, such as ISDN, than an analogue POTS phone. |
| CCAF+ | A CCAF with wireless access functionality. CCAF+ was introduced in the ITU-T CS-2 standard. |
| CCBS | Completion of Calls to Busy Subscribers. The CCBS service is sometimes known as 'ring-back when free'. |
| CCF | Call Control Function. This is an IN construct that models the fundamental call control software in network switches. |
| CCITT | International Consultative Committee on Telephony and Tele-graphy. This was the previous (before June 1994) name for ITU-T. |
| Centrex | The term used for providing a customer with a 'virtual' PBX from a public network switch. |
| CG | CallGap – an INAP operation. INAP operations are described in Section 2.5.1. |
| CI | Correlation Identifier. An ITU-T INAP parameter used for correlating call-processing requests that occur in separate TCAP transactions. |
| CLI | Calling Line Identity. This is sometimes abbreviated to **CgLI**, to distinguish it from the called line identity (**CdLI**). |
| CLIP | Calling Line Identity Presentation. This is a telephony network service that enables customers to display callers' line identities on incoming calls. |
| CLIR | Calling Line Identity Restriction. This telephony service allows the originating end to inhibit the display of calling line identity at the far end, where it over-rides CLIP. |
| COLP | Connected Line Presentation. A telephony service that enables callers to see a display of the terminating line identity when a call is connected. |
| COLR | Connected Line Restriction. A telephony service that allows the terminating end to inhibit the display of terminating line identity at the originator's display, where it over-rides the COLP service if the caller has it enabled. |
| CON | Connect message – an INAP operation. INAP operations are described in Section 2.5.1. Note that the same term is also used in other signalling contexts (besides IN) for different messages. |
| CORBA | Common Object Request Broker Architecture. This is a distributed object technology for client/server middleware that was introduced by the Object Management Group (OMG). |
| Core INAP | *See ETSI Core INAP*. |

| | |
|---|---|
| **CPE** | Customers' Premises Equipment. CPE is telecommunications equipment used in the domestic or business (or other premises) to originate or terminate telecommunications traffic. |
| **CPH** | Call Party Handling. This was the most significant extra IN function package that was included in the ITU-T CS-2 standard, introduced in 1997. It allows call legs of multi-party calls to be manipulated individually under IN service logic control. *See also CVS.* |
| **CRACF** | Call-related Radio Access Control Function. This was introduced in CS-2 to allow IN to be applied to radio and mobility network control. |
| **CS** | Call Segment. A CS is the collection of legs and connection points that represent the transmission and connection resources for an isolated portion of a call. |
| **CSA** | Call Segment Association. A CSA is a coupling of two or more CSs. |
| **CS-1** | Capability Set number 1. This is a generic ITU-T term for the first tranche of a standard. However, in this book CS-1 generally refers to the ITU-T Intelligent Network CS-1 standard. |
| **CS-1R** | CS-1 Refined. ETSI had taken a reduced subset of the original CS-1 release and had included missing detail in order to formulate ETSI Core INAP. This was then fed back into the ITU-T study group and it was released as the ITU-T CS-1R recommendation. |
| **CSE** | CAMEL Service Environment. In a CAMEL GSM system the CAMEL CSE is equivalent to the SCP in a fixed IN architecture. |
| **CSI** | CAMEL Subscription Information (mark). This is in a subscriber's data profile in the HLR and VLR and indicates that the user is a CAMEL subscriber. |
| **CS-n** | Capability Set-n. This refers to ITU-T sequential steps of functional capability. The term is generally applicable to all standards disciplines in the ITU-T, but in this book we nearly always use it to refer to the series of releases of IN standards. In particular, the term **CS-1** is used to refer to the initial milestone series of intelligent networking standards of March 1993. |
| **CTI** | Computer Telephony Integration. |
| **CTR** | ConnectToResource – an INAP operation. INAP operations are described in Section 2.5.1. |
| **CURACF** | Call-Unrelated Radio Access Control Function. *See also CRACF.* |
| **CUSF** | Call-Unrelated Service Function. This is a CS-2 network function that provides access between a user agent (SCUAF) and an SCF for call unrelated transactions such as location updating. It runs a state model called the BCUSM. |
| **CVS** | Connection View State. This is a term used to describe the connection and call-leg configurations that can be modelled by the ITU-T IN CS-2 call party handling (CPH) features. |
| **DDSN** | Digital Derived Services Network. An early AT&T proprietary intelligent network system based on AT&T's earlier number-translation based Direct Services Dialling Capabilities (DSDC). DDSN is sometimes abbreviated to DSN. |

| | |
|---|---|
| **DFC** | DisconnectForwardConnection – an INAP operation. INAP operations are described in Section 2.5.1. |
| **DFP** | Distributed Functional Plane. This is the next plane down from the Global Functional Plane in the ITU-T IN hierarchy. It effectively provides a 'road-map' showing the relationships between the various logical network objects (the functional entities) comprising an intelligent network. |
| **DLE** | Digital Local Exchange. A term that is sometimes used for a traditional PSTN circuit-switching node that has analogue (POTS) and digital (e.g. ISDN) lines serving end customers. |
| **DP** | (IN) Detection Point in the Basic Call State Model. DPs are points, between the PICs (see below) where IN triggers are embedded. For example, there are 17 DPs defined for the ETSI Core INAP BCSM. DPs can be configured as Event or Trigger detection points (E_DPs or T_DPs). |
| **DPC** | Destination Point Code in an SS7 signalling network. |
| **DPE** | Distributed Processing Environment. |
| **DSDC** | Direct Services Dialling Capabilities. This was an early IN development from AT&T in the USA. *See also DDSN.* |
| **DSS1** | Digital Subscriber Signalling System no. 1. This is an ISDN user network interface. It is a common-channel digital access signalling system. The ITU-T recommendation for DSS1 is Q.931 and the ETSI standard is EN 300 403–1. |
| **DTAP** | Direct Transfer Application Part (of SS7), which is carried over TCAP and SCCP and is for call control messages passing between the BSC and MSC in mobile networks. |
| **DTID** | Destination Transaction IDentity, in an SS7 TCAP message. |
| **DTMF** | Dual Tone Multi-Frequency – a tone signalling system available to POTS users with a multi-frequency key-pad. Also known as MF4 signalling. |
| **E_DP** | Event Detection trigger Point. Typically used to trap call-processing events such as 'answer' or 'release'. E_DPs are dynamically armed by the SCF sending RequestReportBCSM messages to the SSF. E_DPs can be active (E_DP-R for Response) or passive (E_DP-N for Notify). |
| **EIA** | Electronics Industry Alliance. This is a body that organises standards development in the USA. |
| **En Bloc** | A mode of telephony signalling (ISUP and TUP) where all address digits are sent in a single call set-up message. The alternative is 'overlap' signalling. |
| **ETC** | EstablishTemporaryConnection – an INAP operation. INAP operations are described in Section 2.5.1. |
| **ETSI** | European Telecommunications Standards Institution. ETSI produces regional standards that are subsequently used in Europe and elsewhere. |
| **ETSI Core INAP** | This is the 1994 ETSI-defined subset of the ITU-T IN CS-1 Standard. It was adopted and ratified as an international standard by the ITU-T in the following year (1995) under the name CS-1R (R for 'Refined'). ETSI Core INAP is a widely used standard. |

| | |
|---|---|
| **EventReportBCSM** | This is an INAP operation that is used by SSPs to respond to RequestReportBCSM operations. INAP operations are described in Section 2.5.1. |
| **FAM** | Final Address Message. This is optionally used for the last address message in a sequence in the SS7 TUP-derived protocols for telephony call control. It is used if the sending node is aware of the number length of the address digit string. See also 'overlap' signalling. |
| **FCC** | (US) Federal Communications Commission. |
| **FCI** | FurnishChargingInformation – an INAP operation. INAP operations are described in Section 2.5.1. |
| **FE** | Functional Entity. Examples of typical intelligent network FEs are the SSF, SRF and the SCF. |
| **FIM** | Feature Interaction Manager. Part of the IN SSF. |
| **FISU** | Fill-In Signalling Units. These are short signalling 'packets' sent by SS7 MTP on an idle signalling link. |
| **GFP** | Global Functional Plane. This is the high-level theoretical description of how service requirements are mapped into network functions in the ITU-T IN standards. |
| **G-MSC** | Gateway MSC. This is the MSC in a mobile network that interfaces to other networks, typically to a PSTN. |
| **GPRS** | General Packet Radio Service. *See also 3G.* |
| **GSL** | Global Service Logic. GSL is the abstract service logic that describes the operation of the SIBs in the IN global functional plane. |
| **GSM** | Originally this was 'Group Speciale Mobile', but the abbreviation was subsequently re-allocated in ETSI to the phrase 'Global System for Mobile communications'. GSM is the European 2nd generation mobile communications standard. |
| **gsmSSP** | GSM Service Switching Point. (The initials 'gsm' before any IN entity denotes the mobile network variant of that entity.) |
| **GT** | Global Title. This is a form of address used in SS7 SCCP routing. GTs do not refer to particular nodes but need to undergo translation (in a GT table – GTT) to resolve to a physically defined address. |
| **GTT** | Global Title Translation – *see GT.* |
| **H. 248** | H.248 is the ITU-T standard for the control of media gateways by media gateway controllers. It is also known as the 'Megaco' protocol. |
| **H. 323** | An ITU-T suite of protocols designed for controlling multimedia communications over IP. |
| **H. 323 Gatekeeper** | The Gatekeeper is a point of 'central' control for service control and provision in H.323 networks. H.323 devices (such as terminals and gateways) register with gatekeepers before they can be given permission (by the gatekeeper) to receive or originate H.323 'calls'. |
| **HLR** | Home Location Register. Every mobile network has at least one HLR to manage the records and account details for the network's customers. |
| **HTML** | HyperText Mark-up Language. |

| | |
|---|---|
| **IAF** | Intelligent Access Function. The IAF is provided in the CS-2 distributed functional plane to allow interconnection between IN and non-IN structured networks. |
| **IAM** | Initial Address Message. This is the initial call set-up message used by the ISUP and TUP protocols in SS7. |
| **i/c** | 'Incoming'. Used as an abbreviation to denote call set-up direction. |
| **ICA** | InitiateCallAttempt – an INAP operation. INAP operations are described in Section 2.5.1. |
| **I_DP** | InitialDP (Detection Point) message – an INAP operation. INAP operations are described in Section 2.5.1. I_DP is usually the first message in a CS-1 IN transaction and it is sent from an SSF to an SCF. |
| **IETF** | The Internet Engineering Task Force. |
| **IF** | Information Flow. Intelligent Network functional entities (FEs) communicate using IFs. Often there is a one-to-one relation between FEs and INAP operations. INAP operations for CS-1 are described in Section 2.5.1. |
| **IFAM** | Initial and Final Address Message. This is an SS7 telephony signalling message, used in SS7 TUP (but not in ISUP), and TUP-derivatives, as an alternative to the IAM for 'en bloc' signalling. |
| **IMSI** | International Mobile Subscriber Identity. This is a series of numerical digits that allows the identification of mobile phones internationally when they are roaming. IMSI uses the ITU-T E.212 numbering plan. |
| **IN** | Intelligent Network. This is a well used anachronism that has been applied to many things, but it usually refers to the introduction of separated service logic into public telephony networks in the manner described in this book. Sometimes the upgraded PSTN is described as an intelligent network itself; sometimes IN refers just to the new network elements. |
| **IN/1 (&2)** | IN/1 and IN/2 were early standards initiatives by Bellcore in the USA. |
| **INA** | Information Networking Architecture. INA was the 1990 Bellcore-inspired forerunner to TINA. |
| **INAP** | Intelligent Network Application Protocol (part of SS7). *See also ETSI Core INAP.* |
| **INCM** | Intelligent Network Conceptual Model. This is the top-down process that was used as the framework for the production of IN standards by the ITU-T. |
| **INWATS** | Inward Wide Area Telecommunications Service. This was a feature of AT&T exchanges in the 1970s that provided an early Freephone service. *See also DDSN.* |
| **IP** | Internet Protocol **or** IN Intelligent Peripheral, depending on the context. |
| **ISDN** | Integrated Services Digital Network. |
| **ISO** | International Standards Organisation. *See also OSI.* |
| **ISP** | Internet Service Provider. |
| **ISUP** | ISDN User Part (of SS7) signalling. |

| | |
|---|---|
| **ITU-T** | International Telecommunications Union – Telecommunications Division. Before June 1994 the ITU-T was known as CCITT. |
| **IUP** | Interconnect User Part (of SS7) signalling IUP is the TUP derivative telephony signalling system that is historically used between public operators in the UK. The preferred standard interconnect protocol is ISUP. |
| **IVR** | Interactive Voice Response. |
| **JAVA API** | The JAVA API for telephony call control (also known as JTAPI) is an extensible API designed for various call control scenarios. |
| **JAIN** | Java for Advanced Intelligent Network. JAIN is a software industry community that specifies APIs for converged PSTN and IP (Internet Protocol) networks. |
| **LEC** | Local Exchange Carrier in the USA. |
| **LNP** | Local Number Portability |
| **Local Exchange** | Same as *DLE*. |
| **LSSU** | Link Status Signalling Unit. This is used to exchange status information between each end of an SS7 signalling link. |
| **MAP** | Mobile Application Part (of SS7). MAP is a well established SS7 user protocol for handling mobility management messages between HLRs, VLRs and MSCs in mobile networks. It is also used between BSCs and MSCs. |
| **MCI** | Malicious Call Interception. A service that is typically implemented using traditional switch-based functions (but nevertheless does appear on the ITU-T CS-1 list of 'benchmark services'). |
| **MDF** | Main Distribution Frame. A termination point for cables and subscriber lines in a traditional telephone exchange. The MDF provides a flexibility point for jumpering cables to exchange equipment. |
| **MF4** | Multi-Frequency number 4. This is an ITU-T standard for in-band dual tone signalling used by push-button phones. Also known as DTMF and sometimes abbreviated to 'MF signalling'. |
| **MGC** | Media Gateway Controller. An MGC is an external controller for media gateways in IP (Internet Protocol) networks. The MGC provides call control functions, and commonly communicates with the media gateways using the H. 248 standard for supervising call channels. |
| **MSRN** | Mobile Station Roaming Number. |
| **MSU** | Message Signalling Unit. The MSU is the 'payload' of an SS7 MTP message. |
| **MTP** | Message Transfer Part of SS7 signalling. |
| **NANP** | North American Numbering Plan. |
| **NAP** | Network Access Point. The NAP is a CS-1 physical entity that just contains CCAF and CCF functions, and so does not communicate with an SCP. |
| **NCP** | Network Control Point, the term used for a central database in the early DSDC system in the USA. *See also DDSN.* |
| **NICC** | Network Interoperability Consultative Committee. This is a UK standards body, part of OFTEL, which oversees the production of PNO specifications for the interworking of public network operating companies in the UK. |

| | |
|---|---|
| **NUP** | National User Part of SS7 signalling. NUP is derived from the SS7 Telephony User Part (TUP). |
| **O_** | Any IN message or trigger point prefixed by 'O_' is to do with the originating (as opposed to terminating) half of the BCSM call model. |
| **O_BCSM** | Originating BSCM. |
| **OCCRUI** | Out-Channel Call-Related User Interaction. This is an ITU-T CS-2 function introduced to allow direct communication between a user's equipment and the controlling service logic in the SCF. |
| **OCUUI** | Out-Channel Call-Unrelated User Interaction. This is an ITU-T CS-2 feature that is modelled by the BCUSM. |
| **o/g** | 'Outgoing'. Used as an abbreviation to denote call set-up direction. |
| **O&A** | Operational and Administration. |
| **OFTEL** | UK Office of Telecommunications. OFTEL is the regulator for the UK telecommunications industry. |
| **OMG** | Object Management Group. The OMG is a powerful software industry consortium of over 800 companies and is a distributed object industry standards group. It is noted for the introduction of CORBA. |
| **OPC** | Originating Point Code in an SS7 signalling network. |
| **OSA** | Open Service Access. OSA is a consortium that takes account of ETSI, JAIN, 3GPP and Parlay Consortium results and is working to harmonise the production of APIs that can be used with ITU-T IN functional components. |
| **OSI** | Open System Interconnect. The OSI Reference Model (7-layer model) was developed in 1978 by the International Standards Organisation (ISO). |
| **OSS** | Operational Support System. This is a generic term encompassing the background management schemes for administering network functions such as service provision, billing, network management, performance monitoring, statistics collection, etc. |
| **OTID** | Originating Transaction IDentity, in an SS7 TCAP message. |
| **Overlap** | A mode of telephony signalling (in SS7 ISUP and TUP) where all address digits are sent in individual set-up messages as digits become available. The series of set-up messages consists of firstly an IAM, then subsequent SAMs and (optionally, in TUP) a FAM. This can lead to faster call set-ups when interworking with slower signalling systems, but with the penalty of less efficient messaging in terms of quantity of separate messages. The more efficient alternative is 'en bloc' signalling. |
| **PA** | Play Announcement – an INAP operation. INAP operations are described in Section 2.5.1. |
| **Parlay Consortium** | The Parlay Group was formed to produce technology-independent APIs for use across multiple network platform environments. Parlay is now part of the OSA initiative. |
| **P&C** | Prompt and Collect User Information – an INAP operation. INAP operations are described in Section 2.5.1. |
| **PBX** | Private (Branch) Exchange. |
| **PCM** | Pulse Code Modulation (for digital transmission). |

| | |
|---|---|
| **PE** | (IN) Physical (network) Entity. |
| **PINT** | PSTN/Internet Interworking. PINT services are characterised by requests originating from the Internet and resulting in the invocation, or the enrichment, of calls in a traditional telephone network. |
| **PIC** | 'Point In Call'. This is a call processing reference point defined for the IN basic call state model (BCSM). |
| **PIN** | Personal Identification Number. PINs are used in IN structured networks to provide users with secure access to their network services. |
| **PNO** | Public Network Operator. *See also NICC.* |
| **PoP** | Point of Presence (for an Internet service provider). |
| **POTS** | Plain Old Telephone System. |
| **PSTN** | Public Switched Telephone Network – sometimes referred to as POTS. |
| **Q** | The identifier for the ITU-T series of recommendations that cover the topics of switching and signalling. |
| **Q.931** | A well known ITU-T standard for access signalling between an ISDN terminal and a narrowband ISDN network. |
| **RBOC** | Regional Bell Operating Company. |
| **RCF** | Radio Control Function. *See also CRACF.* |
| **RCU** | Remote Concentrator Unit. The RCU houses terminations and control equipment of a group, typically of several thousand, of customers' telephone lines. The RCU is connected to the parent exchange over 2Meg links. |
| **RequestReportBCSM** | This is an INAP operation that is used to arm IN E_DPs in the ITU-T IN call model. INAP operations are described in Section 2.5.1. |
| **RIL 3** | GSM Radio Interface Layer 3, which is based on DSS1 and is carried over SS7 DTAP and SCCP. |
| **ROSE** | Remote Operations Service Element. This is the basis of the TCAP component sub-layer and is defined in ITU-T X.229. |
| **RP** | Reference Point (in TINA). |
| **SAD** | Send All Digits. This is a TUP message that can be sent in response to an IAM in order to obtain all remaining address digits. |
| **SAM** | Subsequent Address Message. This is an ISUP (and TUP) message and it follows the IAM. It contains extra address digits. |
| **SA-GF** | Service Application Gateway Function. SA-GF is an IN CS-4 component that provides an open API interface to an IN structured network. |
| **SC-GF** | Service Control Gateway Function. SC-GF is an IN CS-4 component that provides an interface between and IP (Internet Protocol) network node and an IN SCF. |
| **SCCP** | Signalling Connection Control Part (of SS7). This allows non-circuit-related signalling (such as TCAP) to be carried over SS7 routes. |
| **SCE(F)** | Service Creation Environment (Function) in an intelligent network. |

| | |
|---|---|
| **SCF** | Service Control Function. This is most often located in an SCP, but can also be deployed in other IN nodes such as IP, SSP or SN. |
| **SCF Id.** | SCF Identifier. An ITU-T INAP parameter. |
| **SCP** | Service Control Point. |
| **SCUAF** | Service Control User Agent Function. Introduced in IN CS-2. *See also CUSF*. |
| **SDF** | Service Data Function. This is a database function that may be located in an IN physical entity such as an SCP, SN or SSP. It can also reside in a dedicated and separate SDP. |
| **SDP** | Service Data Point. *See also SDF*. |
| **SDL** | Specification Description Language. This is a commonly used diagrammatic method of illustrating and describing the sort of state changes that are typical of IN functions. An example of the use of SDL is the BCSM description in Q.1214. |
| **SFR** | Service Filtering Response – an INAP operation. INAP operations are described in Section 2.5.1. |
| **SHP** | Service Handling Protocol. An SS7 TUP signalling parameter, carried in the IAM or IFAM that informs the far end about the nature of the service that is required, e.g. POTS or ISDN. |
| **SIB** | Service Independent Building block. SIBs are abstract representations of an IN's capabilities and they are intended to provide the fundamentals for a generic capability for building IN-based services. |
| **SIO** | Service Information Octet. SIO contains information on the identity of the user of an SS7 MTP message. SIO is also an abbreviation for 'signalling indication 'out-of-alignment', which is transmitted when one end of an SS7 signalling link is attempting to align with the other end. |
| **SIP** | Session Initiation Protocol (SIP). SIP is an IETF initiative that provides an alternative (to the ITU-T H.323 standard, where Q.931-based call set-up messages are exchanged) method for setting up and modifying conferences between multiple parties in a VoIP scenario. SIP is based on HTML-type techniques. |
| **SIP Proxy** | An SIP Proxy contains both an SIP client and server. It is therefore able to act as a 'go-between' at the functional interface between SIP-based packet networks and traditional IN-structured networks. |
| **SK** | Service Key. SK is a parameter carried in an I_DP that allows a service logic reference to be conveyed from the SSF to the SCF. |
| **SLEE** | Service Logic Execution Environment. This is traditionally the term used for the 'front' end of an SCF. The SLEE controls the transaction management aspects of an IN invocation session, ensuring that IN requests and responses are handled by the correct SLP. |
| **SLP** | Service Logic Program (in an SCF). The SLP represents the service logic for a particular IN service or feature. |
| **SMAF** | Service Management Access Function. SMAF provides the access interface to SMF from service users as well as network operator functions. |
| **SMF** | Service Management Function *(see also TMN)*. |

**SMG**          Special Mobile experts Group – an ETSI working party that invented GSM.

**SMS**          (GSM) Short Message Service **or** (IN) Service Management System.

**SN**           Service Node. This is a self-contained IN network element that typically contains SCF, SSF and SRF entities.

**SND**          Send 'N' Digits. This is a TUP message that can be used as an alternative to SAD if the recipient of the preceding IAM knows how many digits to request. 'N' represents the number of digits requested. The use of SND is more precise and leads to faster call set-ups because the recipient node can proceed with call set-up rather than waiting for possible arrival of extra address digits, as is the case with the SAD procedure.

**SP**           SS7 Signalling Point.

**SPC**          Stored Program Control. This was the term used for early computer-controlled telephone exchanges.

**SPIRITS**      Services In the PSTN/IN Requesting Internet Services. SPIRITS services are the converse of PINT services – events in an intelligent telephony network can instigate actions in a data network. Internet call waiting is an example of such a service.

**SRF**          Specialised Resource Function. The SRF provides speech path resources that are additional to those provided in the CCF. SRFs are typically located in IN IPs or in SSPs.

**SRR**          Specialised Resource Report – an INAP operation. INAP operations are described in Section 2.5.1.

**SSCP**         Service Switching Control Point. This is a composite node, encompassing the IN functions typically contained by SSPs and SCPs.

**SSF**          Service Switching Function. The SSF is distinguished from the SSP in that the latter term refers to a physical network node containing an SSF, possibly in addition to other IN functions.

**SSN**          SCCP Sub-system Number. SSN is used in non-circuit-related SS7 signalling to represent the SCCP user.

**SSP**          Service Switching Point – generally refers to a traditional voice network switch that has been enhanced with IN functions such as the ability to 'trigger' to external service logic when the trigger conditions are met.

**STP**          SS7 Signal Transfer Point.

**T_**           Any IN message or trigger point prefixed by 'T_' is to do with the Terminating (as opposed to originating) half of the BCSM call model.

**T_BCSM**       Terminating BSCM.

**TC**           Transaction Capabilities. *See also TCAP.*

**TCAP**         Transaction Capabilities Application Part. TCAP is an SS7 user part that provides a remote-operations transaction service to higher level signalling protocols, such as INAP. TCAP uses the underlying SS7 transport services of SCCP and MTP.

**T_DP**         Trigger Detection Point. T_DPs can be active (T_DP-R for response) or passive (T_DP-N for Notify).

| | |
|---|---|
| **Teletex** | An ITU-T defined international store and forward data service for error free communications over the switched telephone network. |
| **TIA** | Telecommunications Industry Association (a US standards authority). |
| **TINA** | Telecommunications Information Network Architecture. TINA was a 7-year project that was concerned with the introduction of DPE solutions and products into the telecommunications arena. |
| **TMN** | Telecommunications Management Network. TMN is an ITU-T standard for management functions that is used by IN CS-2. The ITU-T TMN standards are M.3010 and M.3400. |
| **TUP** | Telephony User Part (of SS7) signalling. |
| **UFM** | Unified Functional Methodology. This is a CS-2 mechanism for managing feature interactions. |
| **UMTS** | Universal Mobile Telecommunications Service. UMTS provides the European standard for 3G, specified in ETSI GSM Release '99. Because of the prominence of ETSI GSM in mobile standards UMTS is set to become a major worldwide standard for 3G mobile users (see also '3GPP' above). Whereas GSM uses TDMA (Time Division Multiple Access), where each device has a fixed time-slot allocation for the duration of a call, UMTS uses W-CDMA (Wide-band Code Division Multiple Access). The UMTS frequency bands are 1920–1980 MHz for the upward direction and 2110–2170 MHz for the downward direction. |
| **UNI** | User Network Interface. |
| **Videotex** | An ISDN service that delivers textual information via the phone line (or cable) to a TV set. This is a service that did not achieve any significant popularity. |
| **VLR** | Visitors' Location Register. This is associated with an MSC and holds temporary identity information about mobiles which are currently in the MSC's domain. |
| **VoIP** | Voice over Internet Protocol. |
| **VPN** | Virtual Private Network. This appears to the users to be a private network that uses public network resources. |
| **VRU** | Voice Response Unit. Commonly provided as an IN IP node. |
| **WIN** | Wireless Intelligent Network. This is the IN standard for ANSI-41 based networks in the same way that CAMEL is the accepted standard for the introduction of IN features to GSM networks. |

# Index